SPECIAL PAPERS IN PALAEONTOLOGY NO. 64

INOCERAMID FAUNAS AND BIOSTRATIGRAPHY OF THE UPPER TURONIAN–LOWER CONIACIAN OF THE WESTERN INTERIOR OF THE UNITED STATES

BY

IRENEUSZ WALASZCZYK *and* WILLIAM A. COBBAN

with 32 plates, 2 tables and 27 text-figures

THE PALAEONTOLOGICAL ASSOCIATION
LONDON

December 2000

CONTENTS

ABSTRACT. Twenty-seven inoceramid species and subspecies of the genera *Inoceramus* J. Sowerby, 1814, *Mytiloides* Brongniart, 1822, *Cremnoceramus* Cox, 1969, and *Tethyoceramus* Sornay, 1980 are described from the Upper Turonian and Lower Coniacian of the Western Interior of the United States. One *Inoceramus* species, *I. dakotensis*, and two *Mytiloides* species, *M. ratonensis*, and *M. bellefourchensis*, are described as new, and four forms are left in open nomenclature. Four widely used North American species, *I. dimidius* White, 1874, *I. perplexus* Whitfield, 1877, *C. deformis erectus* (Meek, 1877), and *C. deformis deformis* (Meek, 1871), are thoroughly discussed for the first time and illustrated. The inoceramids allow a subdivision of the Upper Turonian and Lower Coniacian into eight zones, and their precise correlation with the European succession and the standard substage subdivision of the Turonian and Coniacian, as well as with the ammonite zonation used in the US Western Interior. The base of the Upper Turonian, corresponding to the base of the European *I. costellatus* Zone, corresponds to the base of the *I. perplexus* Zone, and to the base of the *Scaphites whitfieldi* ammonite Zone. The Upper Turonian is divided into zones of *Inoceramus perplexus*, *I. dakotensis*, *Mytiloides incertus*, *M. scupini*, and *Cremnoceramus waltersdorfensis waltersdorfensis*. The Lower Coniacian has the zones of *Cremnoceramus deformis erectus*, *C. deformis dobrogensis*, and *C. crassus crassus*. In the upper part of the *erectus* Zone and in the *dobrogensis* Zone, the *Tethyoceramus wandereri* and *Cremnoceramus crassus inconstans* zones may be distinguished. *Cremnoceramus deformis erectus* (Meek) is the basal member of the *deformis* lineage and the proper name for forms referred to *C. rotundatus* (*sensu* Tröger 1967 *non* Fiege 1930), and, accordingly, is the basal boundary marker of the Coniacian. So defined, the first *Forresteria* have already appeared in the topmost Turonian. The Turonian/Coniacian boundary is marked by a series of short-lived events better recognized in Europe (*Didymotis* I Event, *waltersdorfensis* Event, *herbichi* Event, *erectus* I Event).

THIS report provides the taxonomic description and biostratigraphical evaluation of the Upper Turonian–Lower Coniacian inoceramids from the US Western Interior (Text-fig. 1). Based on selected measured sections and numerous additional collections, our work has revealed a clearly defined inoceramid succession that is the basis for a sequence of inoceramid zones, and a good correlation with the European inoceramid succession. Faunas are somewhat disparate in the lower Upper Turonian, but satisfactory correlation is still possible. Beginning with the upper Turonian *Mytiloides scupini* Zone, inoceramid faunas are almost identical in the two continents, allowing the application of the same inoceramid zonation.

Many of the apparent differences between the Upper Turonian–Lower Coniacian inoceramid faunas of Europe and North America result from variable concepts of many of the inoceramid species that are common to both areas. Inoceramids were initially treated separately in the two areas, and even Seitz (1959) failed to recognize a significant number of common forms between European and North American inoceramid faunas; the exception was *I. deformis*, which he treated as a junior synonym of the European *I. schloenbachi* Böhm 1909. Beginning in the latest 1960s and in subsequent years Kauffman (1975, 1978*a*, *c*; Kauffman *et al.* 1978, 1993) argued strongly for close similarities between the inoceramid faunas of the two regions. He applied, however, different taxonomic concepts to many species when compared to previous authors (or used open nomenclature), and this created serious problems for other workers, particularly when attempting precise correlations between the European and North American inoceramid successions based on Kauffman's stratigraphical charts. Many well-known European species appear in different stratigraphical positions in North America and in Europe according to his data. The lack of supporting taxonomic accounts has made it impossible to confirm Kauffman's results.

Furthermore, such American forms as *Inoceramus erectus* Meek, 1877, and *Inoceramus perplexus* Whitfield, 1877, both well represented throughout the Euramerican Region, were misidentified or referred to other species in Europe as a result of a lack of revision of US faunas.

STAGE AND SUBSTAGE DEFINITIONS

Middle/Upper Turonian boundary

No formal proposal for defining the base of the Upper Turonian substage was submitted to the Brussels Symposium on Cretaceous Stage Boundaries (Rawson *et al.* 1996). An informal working group of inoceramid workers led by Dr K. A. Tröger (Freiberg, Germany), was set up with the purpose of finding a suitable inoceramid datum close to the FOs (first occurrences) of *Romaniceras deverianum* and

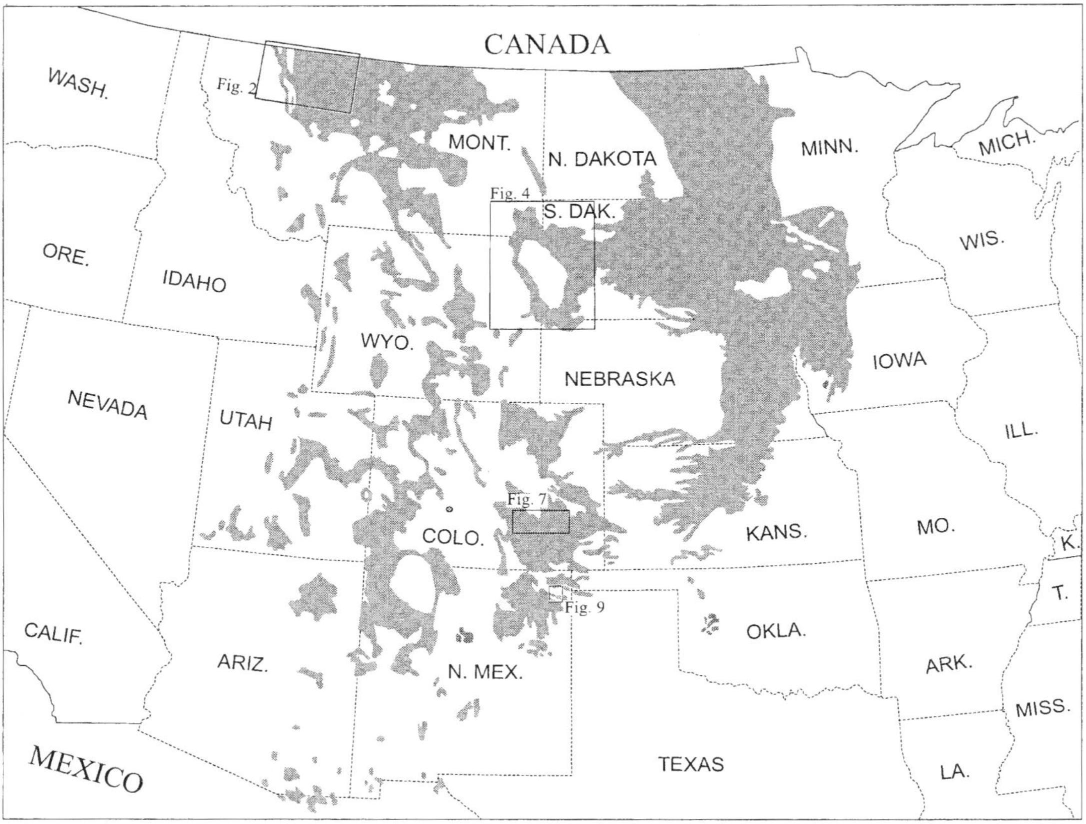

TEXT-FIG. 1. Sketch-map showing the outcrop area of Cretaceous rocks in the Western Interior of the United States; modified from Cobban and Reeside (1952b).

Prionocyclus neptuni. The FO of *Inoceramus costellatus* Woods, 1912, was one of the possibilities discussed, while there has been an assumption on the part of workers in continental Europe that the base of the Upper Turonian lay at the base of a *Subprionocyclus neptuni* ammonite Zone, equivalent to the base of the *Inoceramus costellatus* inoceramid Zone. This is the definition adopted here. Moreover, subsequent taxonomic work has revealed that the *Inoceramus costellatus* of authors is not the same as *Inoceramus costellatus* of Woods, 1912, the holotype of which (Woods 1912, pl. 54, fig. 5) is a *Mytiloides* (Walaszczyk and Wood 1999*a*, p. 428). The abundant material referred to this species by other workers is a junior synonym of *Inoceramus perplexus* Whitfield, 1877. Accordingly, the *costellatus* Zone should be called the *perplexus* Zone.

Turonian/Coniacian boundary

Following the Brussels meeting, the Coniacian Working Group of the Cretaceous Subcommission on Stratigraphy recommended that the FAD (first appearance datum) of *Cremnoceramus rotundatus* (*sensu* Tröger, 1967 *non* Fiege, 1930), the form now shown to be a junior synonym of *Cremnoceramus deformis erectus* (Meek, 1877), as the criterion for recognizing the base of the Coniacian Stage. The goal was to locate the Turonian/Coniacian boundary in the well-represented inoceramid record which, at the same time, would fall in an interval between the horizon with last 'typically Turonian' and the first 'typically

Coniacian' ammonite species (Kauffman *et al.* 1996). Such a coincidence was to be observed in North American record (Kauffman 1995), and was also postulated for the European record where the entrance level of *C. deformis erectus* post-dated the last occurrence of *Prionocyclus germari* and appeared well below the first *Forresteria*, the oldest record of which was known from the base of the *inconstans* Zone (Kaplan 1986; Kaplan and Kennedy 1994, 1996; Kauffman *et al.* 1996). However, in the Western Interior *Forresteria* (*Forresteria*) *peruana* (Brüggen) and *F.* (*Forresteria*) *brancoi* (Solger) appear distinctly below the entrance level of *C. deformis erectus* (Meek), already in the topmost part of the *M. scupini* Zone (see also Kennedy and Cobban 1991). Moreover, *Forresteria* (*Harleites*) cf. *petrocoriensis* (Coquand, 1859) was also recently found in the Upper Turonian *M. scupini* Zone in the Vistula section of central Poland (determination by W. J. Kennedy, 1999).

The requirements of the boundary position discussed in Brussels, i.e. at the base of *C. rotundatus* (=*deformis erectus*) Zone, above the last *Prionocyclus germari* and below the first *Forresteria peruana* and *F. brancoi*, cannot be fulfilled. Taking into account the identity of the inoceramid record in the Western Interior and in Europe, with the same succession of events across the boundary interval (see Text-figs 8, 10), the inoceramids are evidently a more suitable boundary marker than the ammonites, as was proposed in Brussels. The part of the final statement referring to ammonites should, therefore, be excluded from the definition of the Turonian/Coniacian boundary (Walaszczyk and Cobban 1999).

Lower/Middle Coniacian boundary

No internationally consistent Lower/Middle Coniacian substage boundary was agreed during the Brussels meeting. The Coniacian Working Group recommended, however, the FAD of the inoceramid genus *Volviceramus*, particularly the species *Volviceramus koeneni* (Müller, 1888), as the criterion for defining the base of the Middle Coniacian (Kauffman *et al.* 1996, p. 86). Although this part of the succession was not studied in detail during the present study, the collections housed in the Geological Survey in Denver, as well as the material from the Pueblo section, indicate that, as in Europe, it is a very distinct level in inoceramid evolution in the Euramerican biogeographical region, with high correlation potential. The recent study of a very expanded Coniacian succession in the Stafhorst IV mine section, northern Germany (Niebuhr *et al.* 1999), showed, moreover, that volviceramids do not follow directly the Lower Coniacian *Cremnoceramus* fauna, as hitherto assumed, but are separated from the latter by an interval with another distinct fauna, composed in Europe of sulcate representatives of the *lamarcki* group and referred provisionally to *Inoceramus gibbosus* Schlüter, 1877.

EVENT STRATIGRAPHY IN THE TURONIAN/CONIACIAN BOUNDARY INTERVAL

Event-stratigraphy, as a practical method of refined chronostratigraphical correlation based on relatively short-lived isochronous events, was worked out and first applied in northern Germany to correlate the Cenomanian through Coniacian successions in Lower Saxony and Westphalia (Ernst *et al.* 1983). A series of events, classified as eustato-, eco-, tephro-, litho-, phylo-, and oxic/anoxic events, was recognized, with some of them (e.g. eustatoevents) of global correlation potential. Subsequently, a detailed event-succession through the Upper Turonian and Lower Coniacian of Lower Saxony was published by Wood *et al.* (1984). Of a number of ecoevents spanning that interval, most appeared subsequently to be recognizable also in England (Wood *et al.* 1984), Spain (Küchler and Ernst 1989; Küchler 1998; Wiese 1987), Czech Republic (Čech 1989), Poland (Walaszczyk 1992), and as far as western Kazakhstan (Marcinowski *et al.* 1996). The Turonian/Coniacian boundary interval is particularly well defined in that scheme (Wood *et al.* 1984) with at least ten ecoevents characterizing the topmost Turonian and lowermost Coniacian succession. Of these the most widely traced are: *Didymotis* I Event, *Mytiloides herbichi* Event, *Cremnoceramus waltersdorfensis waltersdorfensis* I Event, *Didymotis* II Event, *Cremnoceramus walters-dorfensis waltersdorfensis* II Event, *Cremnoceramus deformis erectus* I Event, and *Cremnoceramus deformis erectus* II Event (Walaszczyk and Wood 1999).

The base of the Upper Turonian, as discussed above, is well marked by the *costellatus* (of authors)/*plana* Event (Ernst *et al.* 1983; Wood *et al.* 1984).

HISTORICAL BACKGROUND

St Louis, at the junction of the Missouri and Mississippi rivers, was the embarking point in the early 1800s for travel to the western territories by trappers and explorers, who followed the Missouri River to its origin in what is now south-western Montana. Travel was at first by canoe and other small craft and later in the mid-1800s by steamboat. Among the explorers were F. B. Meek and F. V. Hayden, who in 1853 went up the Missouri as far as Fort Pierre and then travelled westward overland to the White River badlands (Oligocene on Upper Cretaceous) in present south-western South Dakota. This was Meek's only trip to Nebraska Territory, but Hayden led many expeditions into this area sponsored by the US Army. Meek and Hayden collaborated on many papers from 1856 through 1865. At first they divided the Cretaceous rocks into five numbered units, but in 1861 they gave names to these units (Meek and Hayden 1861, p. 419). The sequence, from oldest to youngest, was as follows: Formation No. 1, Dakota group; Formation No. 2, Fort Benton group; Formation No. 3, Niobrara division; Formation No. 4, Fort Pierre group; Formation No. 5, Fox Hills group.

The Dakota group, named for outcrops near Dakota City, crops out in eastern Nebraska, where it consists mostly of nonmarine sandstone and variegated mudstone. The Fort Benton group was named for the dark marine shale that forms bluffs along the Missouri River valley at Fort Benton in north-central Montana. The Niobrara division was named for white- to yellow-weathering marls that form bluffs along the Missouri River near the mouth of the Niobrara River. Dark marine shale overlying the Niobrara was named for Fort Pierre in the center of present South Dakota. Shallow-water marine sandstone overlying the Fort Pierre was named the Fox Hills group for the Fox Hills in present north-central South Dakota. The word 'Fort' was later dropped from both the Benton and Pierre, which were then referred to as 'shale' or 'formation'.

At a later date Hayden (1876, p. 45) regarded the Benton, Niobrara and Pierre as a single group under the name Colorado group. White (1878, pp. 21–22, 30), however, restricted Colorado to the Benton and Niobrara, and that assignment is still followed. For many years fossils collected in the Western Interior were often reported as being of Colorado age, Benton age, or Niobrara age.

Gilbert (1896) divided the Benton group into three formations, in ascending order: Graneros Shale, Greenhorn Limestone, and Carlile Shale, all named for localities in the Arkansas River drainage in south-central Colorado and not far from the city of Pueblo. He noted the presence of either yellowish sandstone or purplish limestone at the top of the Carlile Shale, but he did not name them.

Logan (1897) gave the name Blue Hill shale to dark shale forming the upper part of the Benton group in Kansas. That unit was later considered the upper member of the Carlile Shale in central and western Kansas (Rubey and Bass 1925; Bass 1926*a*, *b*), and a sandstone unit at the top was named the Codell Sandstone Bed (Bass 1926*b*). The Codell was elevated to member rank by Mather *et al.* (1928), who extended the name geographically to northern Colorado. Kansas terminology for the Carlile Shale was also extended into Colorado in the Arkansas River drainage by Dane *et al.* (1937). The names Fairport, Blue Hill, and Codell, as members of the Carlile Shale, are now widely used over much of eastern Colorado, north-eastern New Mexico, Kansas, Nebraska, and South Dakota. The Blue Hill is especially widespread, and can be recognized as a member of the Mancos Shale in western Colorado and across much of eastern and central Utah, and as an unnamed unit in central and north-western Montana (Cobban 1951*b*). The 'Purplish limestone', noted by Gilbert (1896) in south-central Colorado lying at the top of the Carlile Shale, was given the name Juana Lopez Sandstone Member by Rankin (1944, p. 12) for outcrops on the Mesita Juana Lopez Grant in north-central New Mexico. Dane *et al.* (1966) enlarged the scope of the member to include dark noncalcareous shale with thin beds of sandstone and calcarenite. Juana Lopez is now a widely used name as a member of the Carlile Shale in eastern Colorado, south-western Kansas, and north-eastern New Mexico, and as a member of the Mancos Shale in western Colorado, eastern Utah, and north-western New Mexico.

Before the 1920s Cretaceous rocks and their fossils in the Western Interior region were usually referred to the 'standard' Western Interior groups or formations for age assignment, such as being of Benton or Niobrara age. European Cretaceous stage names were seldom mentioned. Reeside (1923, 1927, 1929, 1930 and later papers) seems to have been the first person to assign the Carlile Shale to the Turonian and the Niobrara to the Coniacian, and to recognize Turonian and Coniacian fossils in the Mancos Shale and equivalent strata.

LITHOSTRATIGRAPHY OF THE UPPER TURONIAN–CONIACIAN INTERVAL IN THE WESTERN INTERIOR

In the central and eastern parts of the Western Interior of the United States (Text-fig. 1), Upper Turonian inoceramid faunas are present in the upper part of the Carlile Shale and, locally, in the lower part of the overlying Niobrara Formation. Lower Coniacian inoceramid faunas are present in the lower part of the Niobrara (Fort Hays Limestone Member and in the lower part of the overlying Smoky Hill Member).

Along the flanks of the Black Hills uplift in western South Dakota, south-eastern Montana, and eastern Wyoming, the Carlile Shale (140–200 m) consists of the following members in ascending order: Pool Creek, Turner, and Sage Breaks (Cobban 1951*b*; Knechtel and Patterson 1962; Robinson *et al.* 1964). The Pool Creek consists of a lower part with grey calcareous concretions and an upper part with reddish ferruginous concretions. The Turner is a silty to sandy shale unit with some thin beds of sandstone and grey to yellow silty concretions. Dark shale with grey, septarian, calcareous concretions make up the Sage Breaks Member.

Southward, away from the Black Hills, the Pool Creek Member is replaced by two units, a lower Fairport Chalk Member and an upper noncalcareous Blue Hill Shale Member with reddish ferruginous concretions (Hattin 1962), both deriving their names from localities in Kansas. The Turner Sandy Member grades southward into calcareous shale to which the name Montezuma Valley Member (Leckie *et al.* 1997) is herein applied. Rocks of Sage Breaks age are usually absent owing to an erosional surface at the base of the Niobrara. North-westward from the Black Hills, the members of the Carlile Shale Formation become less distinct and merge into the Marias River Shale in central and north-central Montana. The Marias River Shale is the upper part of the Colorado Shale of former usage (Cobban 1951*b*). Westward and south-westward from the Black Hills, the Pool Creek and Turner members become very sandy and merge into the upper part of the Frontier Formation of central Wyoming (Merewether and Cobban 1986). The Sage Breaks, however, continues across much of Wyoming as a lower member of the Cody Shale. In western Colorado, eastern Utah, and northern New Mexico, rocks of Carlile age are represented by units of Mancos Shale intertongued with near-shore and nonmarine formations, such as the Tres Hermanos Formation (Hook *et al.* 1983) and Gallup Sandstone of New Mexico (Nummedal and Molenaar 1995) and Ferron Sandstone of Utah (Gardner and Cross 1994).

The Niobrara Formation consists of a lower Fort Hays Limestone Member (9–24 m) and a much thicker Smoky Hill Chalk (or Marl) Member (45–275 m). Owing to facies changes and disconformities, the Fort Hays Member is entirely Turonian in places, entirely Coniacian in places, or both upper Turonian and lower Coniacian (Text-fig. 2). The member is widely distributed in central and western Kansas, western Nebraska, south-eastern Wyoming, and over much of Colorado and parts of northern New Mexico (Longman *et al.* 1998, fig. 10). In part of north-western New Mexico, the Fort Hays is replaced by a thin glauconitic sandstone unit, the Cooper Arroyo Sandstone Member of the Mancos Shale (Landis and Dane 1967). The Fort Hays Member is missing by facies changes and by pre-Late Coniacian erosion over nearly all of the Black Hills area, but west of the Black Hills, in central and western Wyoming, near-shore sandstone beds at the top of the Frontier Formation contain inoceramids of Early Coniacian age (Cobban and Reeside 1952*a*).

LOCALITY DETAILS

US Geological Survey Mesozoic localities referred to in this report are given below. Numbers preceding the USGS number refer to location numbers in Text-figure 11. Prefix D indicates Denver Mesozoic locality numbers; others are Washington, DC, Mesozoic locality numbers.

1. D4481. Shelby in the SE ¼ sec. 16, T. 32 N, R. 2 W, Toole County, Montana. Marias River Shale, from calcareous and ferruginous concretions 41 m below 'K bed' of Cobban *et al.* (1976).

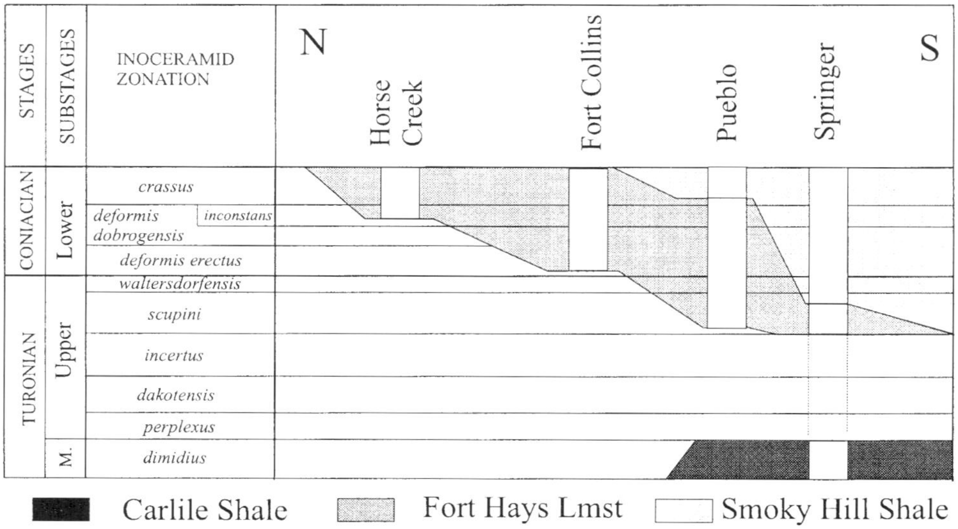

TEXT-FIG. 2. Chronostratigraphical distribution of the Fort Hays Limestone Member of the Niobrara Formation in Wyoming, Colorado and New Mexico.

2. D4482. Same locality as D4481. Marias River Shale, from calcareous concretions 38·5 m below 'K bed'.
3. D4483. Same locality as D4481. Marias River Shale, 37·2–38·4 m below 'K bed'.
4. D4484. Same locality as D4481. Marias River Shale, from ferruginous concretions 35·5–37·0 m below 'K bed'.
5. D4485. Same locality as D4481. Marias River Shale, from 34·4–35·5 m below 'K bed'.
6. D4486. Shelby in the SW ¼ sec. 21, T. 32 N, R. 2 N, Toole County, Montana. Marias River Shale, from ferruginous concretions 33·5 m below 'K bed'.
7. D4487. Same locality as D4486. Marias River Shale, from calcareous concretions 29–33 m below 'K bed'.
8. D4488. Same locality as D4486. Marias River Shale, from ferruginous concretions 28·5 m below 'K bed'.
9. D4489. Same locality as D4486. Marias River Shale, from ferruginous concretions 25 m below 'K bed'.
10. D4490. Same locality as D4486. Marias River Shale, from 22 m below 'K bed'.
11. D4491. Same locality as D4486. Marias River Shale, 15·5–21 m below 'K bed'.
12. D4492. Same locality as D4486. Marias River Shale, from calcareous concretions 12·5 m below 'K bed'.
13. D4493. Same locality as D4486. Marias River Shale, from calcareous concretions 5·5 m below 'K bed'.
14. D4494. Same locality as D4486. Marias River Shale, from septarian concretions 1·8 m below 'K bed'.
15. D4495. Same locality as D4486. Marias River Shale, from the 'K bed' of Cobban *et al.* (1976).
16. 21421. Marias River 8·8 km south of Shelby, Toole County, Montana. Marias River Shale, from 190 m below top.
17. 21422. Marias River valley south of Shelby in the NE ¼ sec. 20, T. 31 N, R. 2 W, Toole County, Montana. Marias River Shale, 160 m below top of Kevin Member.
18. 23280. South-south-east of Shelby in the SW ¼ sec. 8, T. 30 N, R. 1 W, Toole County, Montana. Marias River Shale, from lower part of Kevin Member.
19. 21372. North of Fort Shaw in sec. 35, T. 21 N, R. 2 W, Cascade County, Montana. Marias River Shale, from calcareous concretions 54 m above top of Cone Calcareous Member.
20. 21373. Same locality as 21372. Marias River Shale, from calcareous concretions 0·6–2·1 m above 21372.
21. D12426. SW ¼ sec. 18, T. 3 S, R. 8 E, Park County, Montana. Cody Shale, lower part.
22. 20939. SW ¼ sec. 36, T. 6 S, R. 32 E, Big Horn County, Montana. Carlile Shale, from calcareous concretions 55–56 m above base.
23. 21187. North of Belle Fourche in the SW ¼ sec. 11, T. 9 N, R. 2 E, Butte County, South Dakota. From sandstone concretions 34–35 m above base of Carlile Shale.
24. 21188. Same locality as 21187. Carlile Shale from calcareous concretions 11 m above base of the Turner Sandy Member.
25. 21189. North of Belle Fourche in the N ½ sec. 10, T. 9 N, R. 2 E, Butte County, South Dakota. From calcareous concretions 50 m above base of Carlile Shale.

26. 21190. Same locality as 21189. From ferruginous concretions 59–61 m above base of Carlile Shale.
27. 21192. Same locality as 21189. From ferruginous concretions 73–76 m above base of Carlile Shale.
28. 21194. Same locality as 21189. From ferruginous concretions 76–80 m above base of Carlile Shale.
29. 21195. North of Belle Fourche in the NW ¼ sec. 10, T. 9 N, R. 2E, Butte County, South Dakota. From calcareous concretions 83 m above base of Carlile Shale.
30. 21196. Same locality as 21195. From calcareous concretions 83–86 m above base of Carlile Shale.
31. 21197. Same locality as 21195. From calcareous concretions 86 m above base of Carlile Shale.
32. 21198. Same locality as 21195. From calcareous concretions 90 m above base of Carlile Shale.
33. 21199. Same locality as 21195. From calcareous concretions 101–102 m above base of Carlile Shale.
34. 21765. About 14 km south-south-east of Rapid City in the NE ¼ sec. 22, T. 1 S, R. 8 W, Pennington County, South Dakota. Carlile Shale, from septarian limestone concretions 13–14 m above base of Sage Breaks Member.
35. D8644. North of Moorcroft in the SE ¼ sec. 33, T. 52 N, R, 67 W, Crook County, Wyoming. Carlile Shale.
36. D8999. Oregon Basin oilfield, south-east of Cody, in the SE ¼ sec. 31, T. 51 N, R. 100 W, Park County, Wyoming. Uppermost 4·5–6·0 m of Frontier Formation.
37. D9771. North-west of Thermopolis in the NW ¼ sec. 20, T. 44 N, R. 97 W, Hot Springs County, Wyoming. Frontier Formation, from uppermost part.
38. D9750. North-east of Thermopolis in the NE ¼ sec. 9, T. 43 N, R. 94 W, Hot Springs County, Wyoming. Frontier Formation, from uppermost sandstone bed.
39. D8842. NW ¼ NE ¼ sec. 27, T. 43 N, R. 91 W, Hot Spring County, Wyoming. Cody Shale, from a limestone concretion 21·6 m above base.
40. 10179. Maverick Springs area in T. 6 N, R. 2 W, Fremont County, Wyoming. Uppermost part of Frontier Formation. For stratigraphical section, see Andrews (1944).
41. 3312. Dry Creek, northern part of Shoshone Indian Reservation, in T. 6 N, R. 3 W, Fremont County, Wyoming.
42. 20611. North-west of Lander in the SE ¼ sec. 30, T. 2 N, R. 1 W, Fremont County, Wyoming. Frontier Formation, from a sandstone ledge in upper part.
43. D11939. Winkleman dome north-west of Lander in the NW ¼ sec. 17, T. 2 N, R. 1 W, Fremont County, Wyoming. Frontier Formation.
44. D10458. SE ¼ NE ¼ sec. 14, T. 33 N, R. 94 W, Fremont County, Wyoming. Frontier Formation, Wall Creek Sandstone Member, 3 m below D10459.
45. D10459. SE ¼ NE ¼ sec. 14, T. 33 N, R. 94 W, Fremont County, Wyoming. Frontier Formation, Wall Creek Sandstone Member.
46. 3791. North of Kemmerer in the NE ¼ sec. 13, T. 23 N, R. 116 W, Lincoln County, Wyoming. Frontier Formation.
47. 6773. About 6·4 km north-west of Kaycee, Johnson County, Wyoming. Lower part of Cody Shale.
48. D9118. West of Casper in the NE ¼ sec. 4, T. 33 N, R. 81 W, Natrona County, Wyoming. Frontier Formation, from sandstone concretions in uppermost sandstone.
49. D9357. West of Casper in the NW ¼ sec. 11, T. 33 N, R. 81 W, Natrona County, Wyoming. Frontier Formation, Wall Creek Sandstone Member.
50. D7946. SE ¼ sec. 32, T. 27 N, R. 88 W, Carbon County, Wyoming. Frontier Formation, from basal part of uppermost sandstone.
51. D6625. SE ¼ sec. 24, T. 25 N, R. 85 W, Carbon County, Wyoming. Frontier Formation, from upper part.
52. D7113. North of Rawlins in the NE ¼ sec. 9, T. 22 N, R. 87 W, Carbon County, Wyoming. Frontier Formation, about 60 m below top.
53. D7522. NE ¼ NW ¼ sec. 15, T. 22 N, R. 87 W, Carbon County, Wyoming. Cody Shale, from limestone concretions 22·5 m above base.
54. 22953. East of Rawlins in the SW ¼ sec. 22, T. 21 N, R. 86 W, Carbon County, Wyoming. Frontier Formation, from sandstone concretions 65 m below top. For stratigraphical section, see Cobban and Reeside (1951).
55. D9010. Railroad cut west of Como in the SW ¼ sec. 19, T. 22 N, R. 77 W, Carbon County, Wyoming. Frontier Formation, from upper part of Wall Creek Member.
56. D3763. North of Medicine Bow in SE ¼ sec. 32, T. 23 N, R. 78 W, Carbon County, Wyoming. Frontier Formation, from basal ledge of Wall Creek Sandstone Member.
57. D6928. North-east of Rock River in the SW ¼ sec. 31, T. 22 N, R. 75 W, Albany County, Wyoming. Frontier Formation, from Wall Creek Sandstone Member.
58. D9244. NE ¼ sec. 7, T. 21 N, R. 74 W, Albany County, Wyoming. Frontier Formation, from lowest ledge-forming sandstone bed of Wall Creek Sandstone Member.
59. D8988. SW ¼ NW ¼ sec. 26, T. 17 N, R. 70 W, Laramie County, Wyoming. Niobrara Formation, from limestone bed less than 1 m above base of Fort Hays Limestone Member.

60. 5865. Horse Creek north-west of Cheyenne, Laramie County, Wyoming. Niobrara Formation, from Fort Hays Limestone Member.
61. D1781. SE ¼ sec. 1, T. 8 N, R. 82 W, Jackson County, Colorado. Niobrara Formation, 20 m above the base.
62. D12992. SE ¼ sec. 6, T. 3 N, R. 100 W, Moffat County, Colorado. Mancos Shale, from thin limestone bed 47 m above Frontier Sandstone Member.
63. 23078. Ivie Creek in the NE ¼ SW ¼ sec. 7, T. 23 S, R. 6 E, Sevier County, Utah. Mancos Shale, from limestone concretions about 30 m above base of Blue Gate Member.
64. D3024. South-east of bridge over Currant Creek in the NW ¼ sec. 31, T. 17 S, R. 71 W, Fremont County, Colorado. Niobrara Formation, from chalky limestone bed 1·2 m above base.
65. D3025. Twelve Mile Park in the SE ¼ NW ¼ sec. 31, T. 17 S, R. 17 W, Fremont County, Colorado. Niobrara Formation, from 12–16 m above base of Fort Hays Limestone Member.
66. D13030. NE ¼ NW ¼ sec. 32, T. 20 S, R. 66 W, Pueblo County, Colorado. Niobrara Formation, from upper part of Fort Hays Limestone Member.
67. D7268. Quarry north-east of Lamar in the SW ¼ sec. 19, T. 20 S, R. 45 W, Kiowa County, Colorado. Niobrara Formation, from Fort Hays Limestone Member.
68. D7252. NE ¼ sec. 33, T. 27 S, R. 49 W, Bent County, Colorado. Niobrara Formation, from Fort Hays Limestone Member.
69. D11843. Jack Point in the SW ¼ sec. 35, T. 26 S, R. 58 W, Otero County, Colorado. Niobrara Formation, from basal part of Fort Hays Limestone Member.
70. D11987. Same locality as D11843. Niobrara Formation, from 2·7 m above base of Fort Hays Limestone Member.
71. D2978. Quarry in NE ¼ sec. 33, T. 30 S, R. 60 W, Las Animas County, Colorado. Niobrara Formation, from basal limestone bed of Fort Hays Limestone Member.
72. 18703. North-east of Trinidad in the SE ¼ sec. 35, T. 30 S, R. 60 W, Las Animas County, Colorado. Niobrara Formation, from basal bed of Fort Hays Limestone Member.
73. D9232. Highway 84 south of Pagosa Springs in the SE ¼ sec. 19, T. 35 N, R. 1 W, Archuleta County, Colorado. Mancos Shale, from upper 3 m of shale that underlies the Fort Hays Limestone Member.
74. D3681. South-east of Pagosa Springs in sec. 30, T. 35 N, R. 1 W, Archuleta County, Colorado. Niobrara Formation, basal bed of Fort Hays Limestone Member.
75. D13433. Mill Creek in the NW ¼ SE ¼ sec. 19, T. 35 N, R. 1 W, Archuleta County, Colorado. Niobrara Formation, from limestone bed 1 m below top of Fort Hays Limestone Member.
76. D4395. North of Rio Gallina in the SE ¼ sec. 15, T. 25 N, R. 1 E, Rio Arriba County, New Mexico. Mancos Shale, from basal part of Juana Lopez Member.
77. D11834. North-west of Wagon Mound in the SE ¼ sec. 26, T. 21 N, R. 20 E, Mora County, New Mexico. Fort Hays Limestone, from third limestone bed below top.
78. D12522. North-west of Wagon Mound in the SE ¼ sec. 26, T. 21 N, R. 20 E, Mora County, New Mexico. Niobrara Formation, from uppermost exposed bed of Fort Hays Limestone Member.
79. 8978. 3·2 km south of Sapello, San Miguel County, New Mexico. Niobrara Formation, from Fort Hays Limestone Member?
80. D8284. SW ¼ sec. 22, T. 12 N, R. 1 W, Sandoval County, New Mexico. Gallup Sandstone.
81. 15932. West-north-west of Albuquerque about in the SW ¼ sec. 1, T. 11 N, R. 3 W, Bernalillo County, New Mexico. Basal part of Gallup Sandstone.
82. 3514. Rio Puerco about 5 km north of San Ygnacio, Bernalillo County, New Mexico.
83. 3519. Rio Puerco valley north-west of Albuquerque, Sandoval County, New Mexico.
84. D7002. SW ¼ sec. 2, T. 4 N, R. 11 W, Cibola County, New Mexico. Mancos Shale, from transition beds to overlying Gallup Sandstone.
85. D2494. SW ¼ sec. 1, T. 20 S, R. 4 E, Doña Ana County, New Mexico. Mancos Shale, from calcareous concretions in D-Cross Member. For stratigraphical section, see Lucas and Estep (1998).
86. D12671. NW ¼ sec. 19, T. 20 S, R. 4 E, Doña Ana County, New Mexico. Mancos Shale, from calcareous concretions in D-cross Member. For stratigraphical section, see Lucas and Estep (1998).

MEASURED SECTIONS

Short notes are given below on the most important sections used in this study. They are arranged in geographical order from the Canadian border southwards (indicated by the rectangles on the collections locality map, Text-fig. 11). The location of the maps in Text-figures 3, 5, 7 and 9 are shown on Text-figure 1, the general map of the Western Interior.

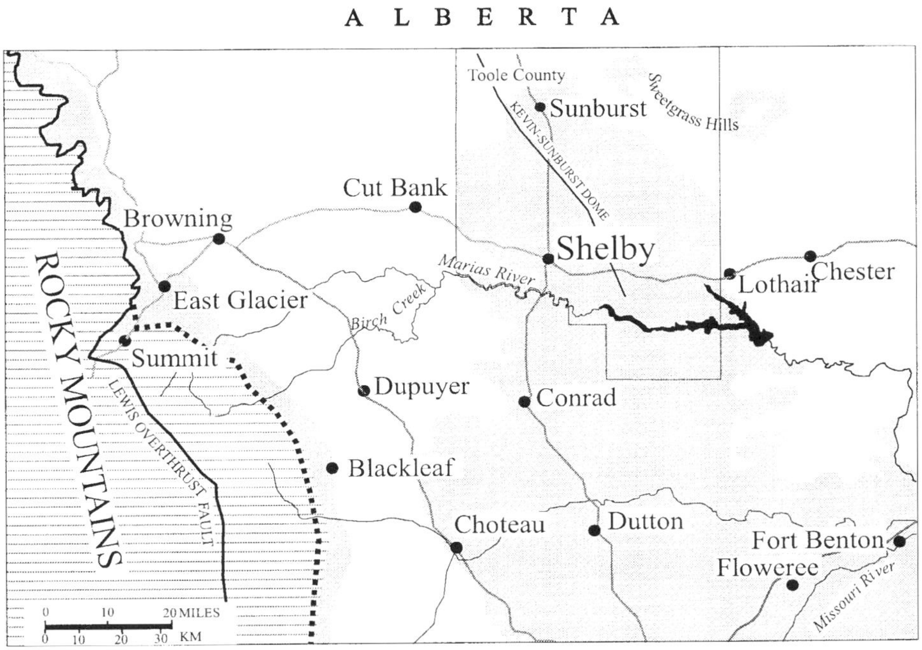

TEXT-FIG. 3. Geological sketch-map of the Shelby area, northern Montana; stippled area marks the Turonian–Santonian outcrop area in the region; see Text-figure 1 for location on the general map of the Western Interior; modified from Cobban *et al.* (1976).

Shelby and Johnson Bridge sections

Both sections span the topmost Turonian–Lower Coniacian succession of the Kevin Member of the Marias River Shale near Shelby, north-western Montana (Text-figs 3–4). Well-preserved inoceramids and ammonites come from the ferruginous and calcareous concretions occurring in several horizons scattered throughout the succession. Of particular importance is the Shelby section, with numerous precisely located collections from the entire Lower Coniacian succession.

North Belle Fourche and Spring Creek sections

These sections, located in the north-eastern and eastern periphery of the Black Hills, South Dakota (Text-fig. 5), include the upper Middle and all of the Upper Turonian of the region with rich inoceramid and ammonite faunas (Text-fig. 6). All of the fossils come from calcareous or ferruginous concretions. The North Belle Fourche section provides rich material from the lower part of the interval studied, spanning the *I. dimidius* through *M. incertus* Zones. The succession can be extended up into the Lower Coniacian on the basis of the data from a borehole in the Osage oilfield, on the western flank of the Black Hills (Cobban 1984*b*).

Pueblo section

An upper Upper Turonian and Lower Coniacian succession, represented by the Fort Hays Limestone Member of the Niobrara Formation and the lower part of the shale and limestone unit of the Smoky Hill Member (Text-fig. 8), is well exposed over a large area of the Rock Canyon anticline, west of Pueblo (Text-fig. 7). The Fort Hays Limestone rests disconformably on the Codell Sandstone or, locally, on a few

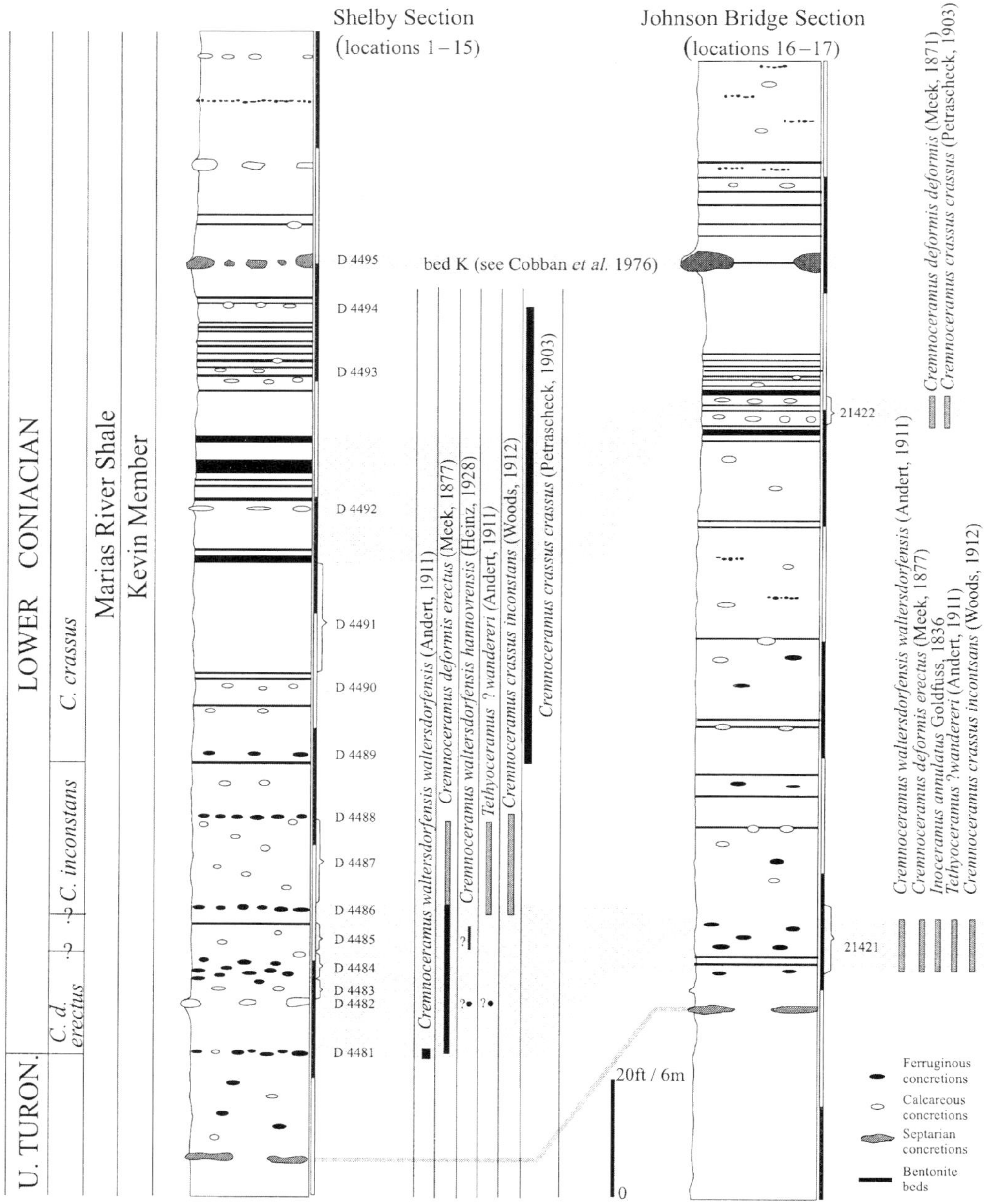

TEXT-FIG. 4. Geological columns, biostratigraphy and inoceramid distribution in the Shelby and Johnson bridge sections, near Shelby, northern Montana. Columns show collection numbers and some bed numbers. Localities D4481 through D4495, 21421, and 21422 (1–17 in locality list) in Text-figure 11.

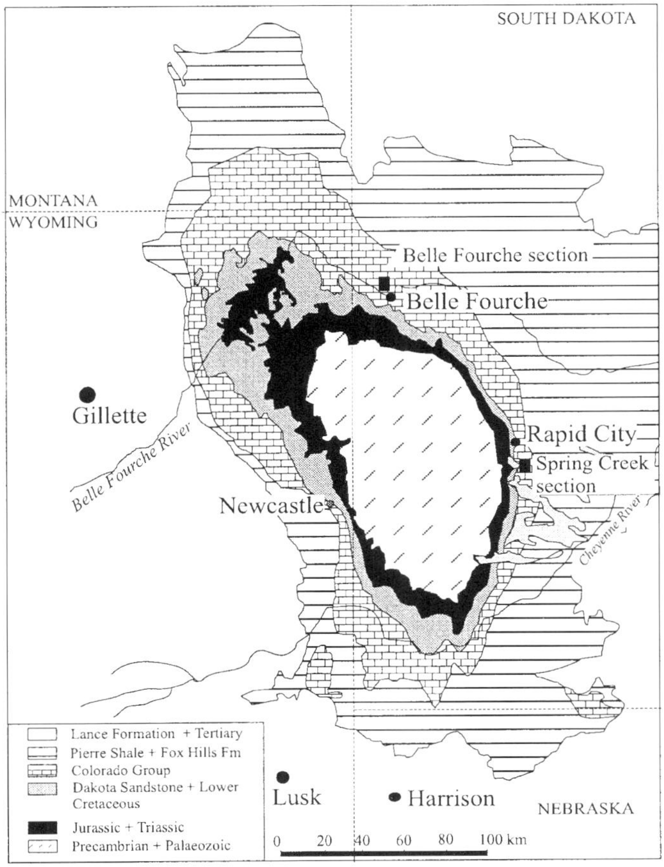

TEXT-FIG. 5. Geological sketch-map of the Black Hills area; see Text-figure 1 for location in the general map of the Western Interior; modified from Stose (1932).

centimetres of dark shale that represents the basal bed of the Juana Lopez Member of the Carlile Shale. In many places clasts of Juana Lopez Limestone, derived from fossiliferous concretions, are present in the basal bed of the Fort Hays (Cobban 1988)

The Upper Turonian succession begins with the inoceramid-rich interval that occurs commonly at the base of the *M. scupini* Zone, and is followed by a 'barren interval' with an extremely rare fauna. Inoceramids are abundant again near the Turonian/Coniacian boundary. In spite of the apparently reduced sedimentation rate in the area, the stratigraphical record of this succession seems to be quite complete (if there is a small gap or condensation, this probably lies at the *erectus/dobrogensis* Zone boundary, as indicated by the co-occurrence of *Cremnoceramus erectus* and rather advanced members of *C. deformis*). The completeness of most of the Pueblo succession is demonstrated by the Turonian/Coniacian boundary interval, comprising a very precise record of the inoceramid assemblages across the boundary (Walaszczyk and Cobban 1999). It comprises an almost complete set of the boundary events, as distinguished originally in Europe (Ernst *et al.* 1983; Wood *et al.* 1984; see also Walaszczyk and Wood 1999*a*, *b*), the *Didymotis* I, *waltersdorfensis*, and *erectus* Events. Moreover, below the *waltersdorfensis* Event a poorly known '*websteri* fauna' occurs.

The other very important feature of the Pueblo section is the ammonite record across the Turonian/

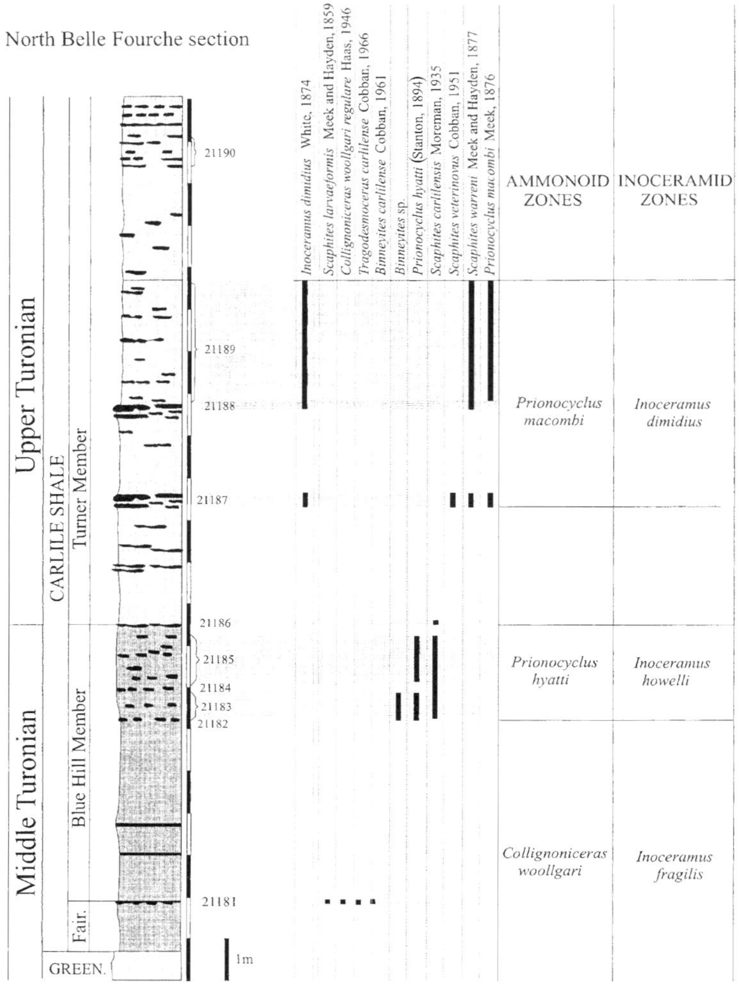

TEXT-FIG. 6. Geological columns, litho- and biostratigraphy, and vertical distribution of the ammonite and inoceramid faunas in the North Belle Fourche and Spring Creek sections; see Text-figure 5 for location. Collection numbers are on right-hand side of columns; some bed numbers from measured section are on left-hand side; Fair, Fairport Member.

Coniacian boundary, with the first *Forresteria* occurring well below the base of the Coniacian, as defined by the first occurrence of *Cremnocerams deformis erectus* (Meek) [=*C. rotundatus* (*sensu* Tröger, 1967 *non* Fiege, 1930)].

The Coniacian part of the Fort Hays Limestone is dominated by representatives of the *Cremnoceramus deformis* lineage: *C. deformis erectus* and *C. deformis dobrogensis*. *C. crassus crassus* (Petrascheck) appears at the top of the Fort Hays and dominates in the lower part of the succeeding shale and limestone unit of the Smoky Hill Member (see Text-fig. 8).

La Junta section

This section represents the same Niobrara interval as that exposed near Pueblo (Text-fig. 8), and is located about 100 km east of the latter (Text-fig. 7). The best exposure is located in the north-east part of the town,

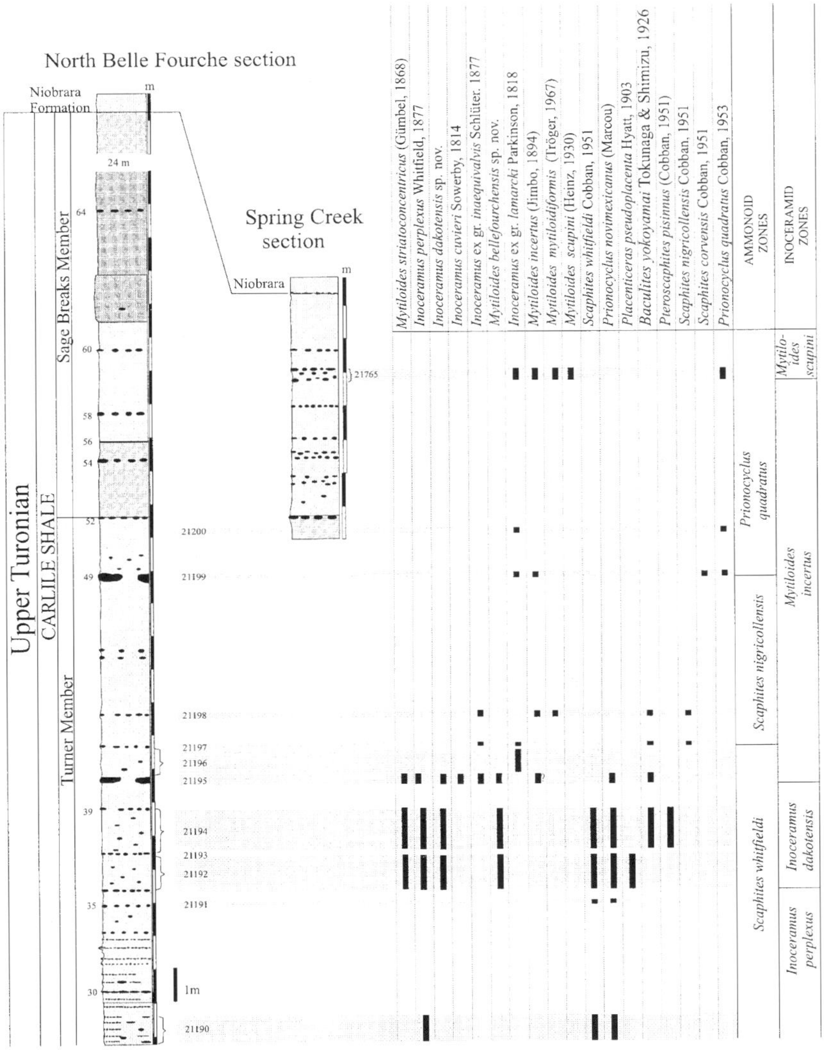

TEXT-FIG. 6. Continued.

along State Highway 109, north of the Arkansas River valley. As at Pueblo, the first specimens of *Forresteria* occur well below the base of the Coniacian as defined by the first occurrence of *C. deformis erectus*.

Springer and Wagon Mound sections

The Upper Turonian Wagon Mound and Upper Turonian–Lower Coniacian Springer sections in the Raton Basin in north-east New Mexico (Text-fig. 9) expose Fort Hays Limestone and the shale and limestone unit of the Smoky Hill Member (Text-fig. 10).

The Fort Hays Limestone in the Raton Basin differs from that in the Pueblo-La Junta area in the higher shale content and in its biostratigraphical range. Its upper boundary is here at the base of the 'barren

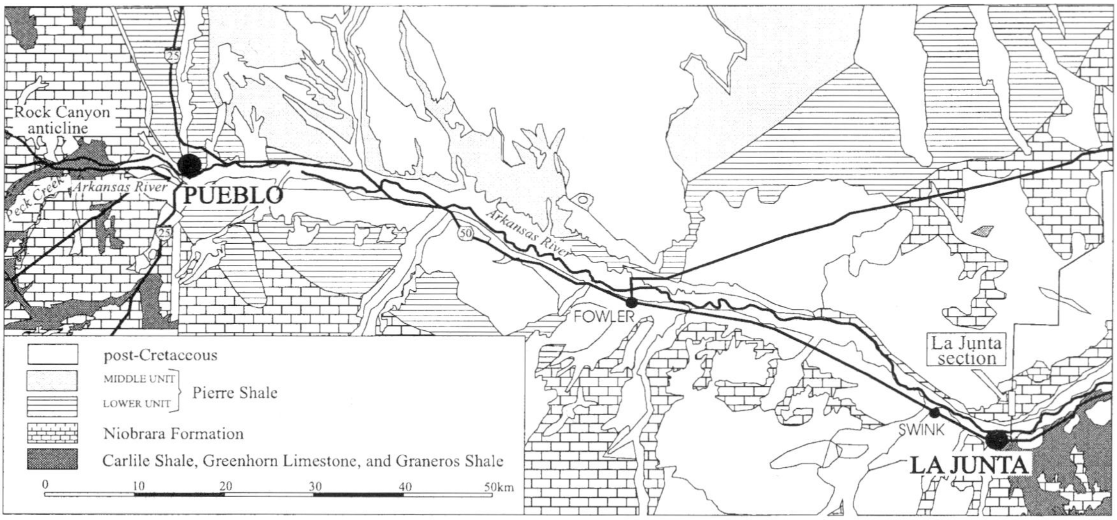

TEXT-FIG. 7. Geological sketch-map of the Pueblo-La Junta area, south-eastern Colorado; see Text-figure 1 for location in the general map of the Western Interior; modified from Tweto (1979).

interval' of the *M. scupini* Zone, and it is well below its upper boundary in the Pueblo-La Junta area, where it ranges to the upper boundary of the *deformis dobrogensis* Zone of the Lower Coniacian. The lower boundary in both areas lies approximately at the same level, at the base of the *M. scupini* Zone, but the upper boundary is substantially older, at the base of the barren interval of the *scupini* Zone at Springer, rather than at the *deformis dobrogensis-crassus* Zone boundary, as is the case at Pueblo (Text-fig. 2).

Both Springer and Wagon Mound sections expose the lower, fossiliferous part of the *M. scupini* Zone, with rich and very well-preserved faunas of late Turonian inoceramids: *M. ratonensis* sp. nov., *M. scupini* (Heinz, 1930), *Mytiloides* sp., and *Inoceramus longealatus* Tröger, 1967 (see also Scott *et al.* 1986, fig. 6a–l). This is succeeded by the barren interval with very scarce fauna. Although the base of the barren interval falls near the facies change in the Springer-Wagon Mound sections, it is a facies-independent phenomenon, also being well developed in the Pueblo-La Junta area, within the Fort Hays Limestone. A similar barren interval is also present in Europe at this level (Ernst *et al.* 1983; Wood *et al.* 1984; Walaszczyk and Wood 1999*a*). The Turonian/Coniacian boundary is marked by the same set of boundary events as in the Pueblo-La Junta area.

The Wagon Mound section (Text-fig. 10) was misinterpreted by Collom (1991) and Kauffman (1995; see also Kauffman *et al.* 1996). The inoceramid-rich beds in the Fort Hays Limestone was taken by these authors to contain the Turonian/Coniacian boundary interval, but this is actually far higher; the section is exclusively Upper Turonian. Moreover, the topmost three beds (LS 8 through LS 10 in Text-fig. 10) already represent the shale and limestone unit and not the Fort Hays Limestone Member as marked on their scheme.

INOCERAMID ZONATION IN THE US WESTERN INTERIOR

Ten inoceramid zones are here recognized in the Upper Turonian–Lower Coniacian substages (Tables 1–2). The topmost Turonian *Cremnoceramus waltersdorfensis waltersdorfensis* Zone and the Lower Coniacian *C. deformis erectus* and *C. crassus crassus* (=*crassus/deformis*) Zones are used here in the sense recently defined by Walaszczyk and Wood (1999*a*), unless indicated otherwise. All zones are interval zones defined by the first occurrence of their index taxon (lowest-occurrence zone; see Salvador 1994) and their correlation with the standard ammonite zonation for the Western Interior (Cobban 1951*b*, 1984*a*;

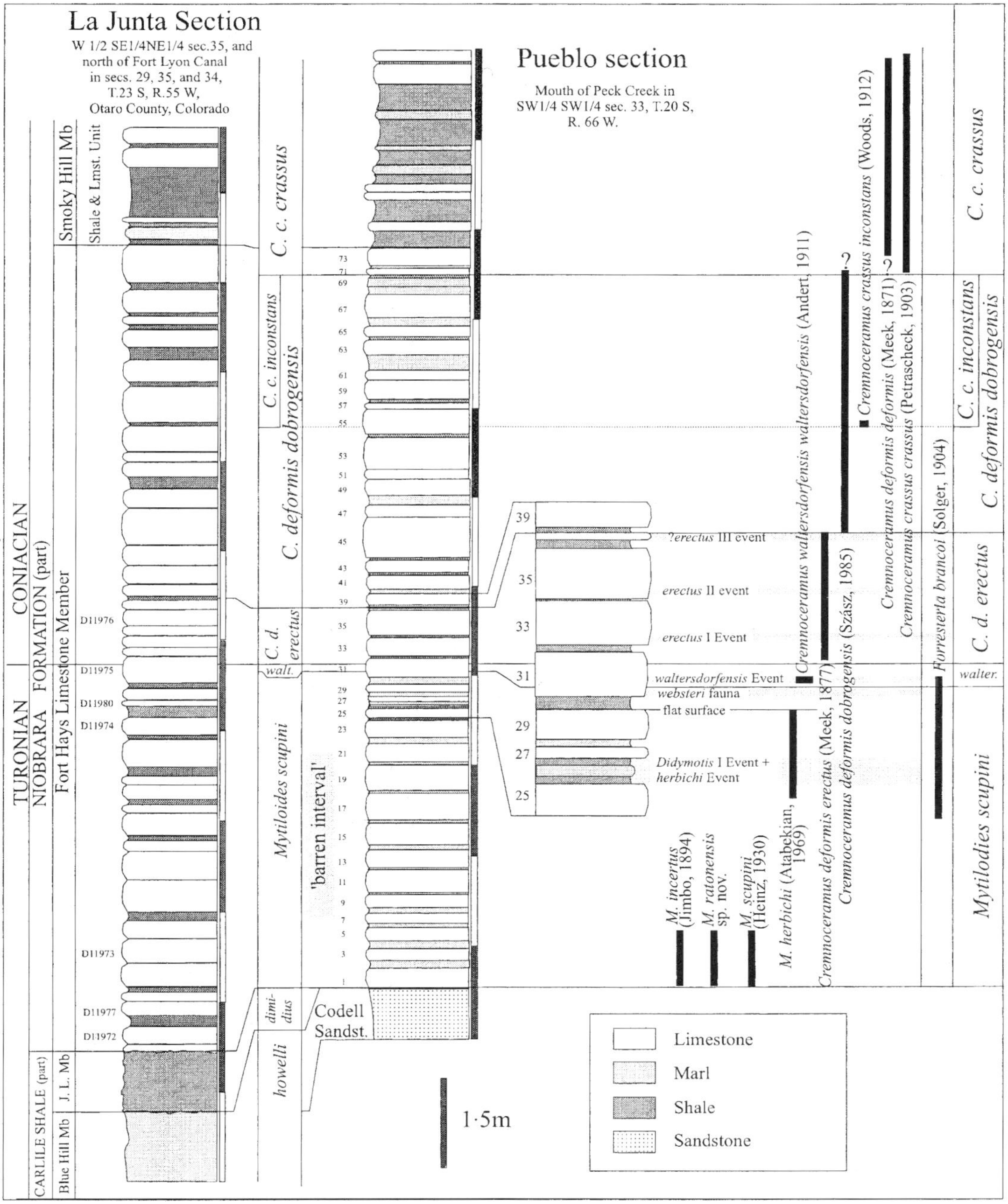

TEXT-FIG. 8. Geological columns, litho- and biostratigraphy, and vertical distribution of inoceramids in the La Junta and Pueblo sections, south-eastern Colorado; see Text-figure 7 for location. Numbers on left-hand side of the Pueblo section are after Scott and Cobban (1964); J.l., Juana Lopez Member; *walter*, *waltersdorfensis*.

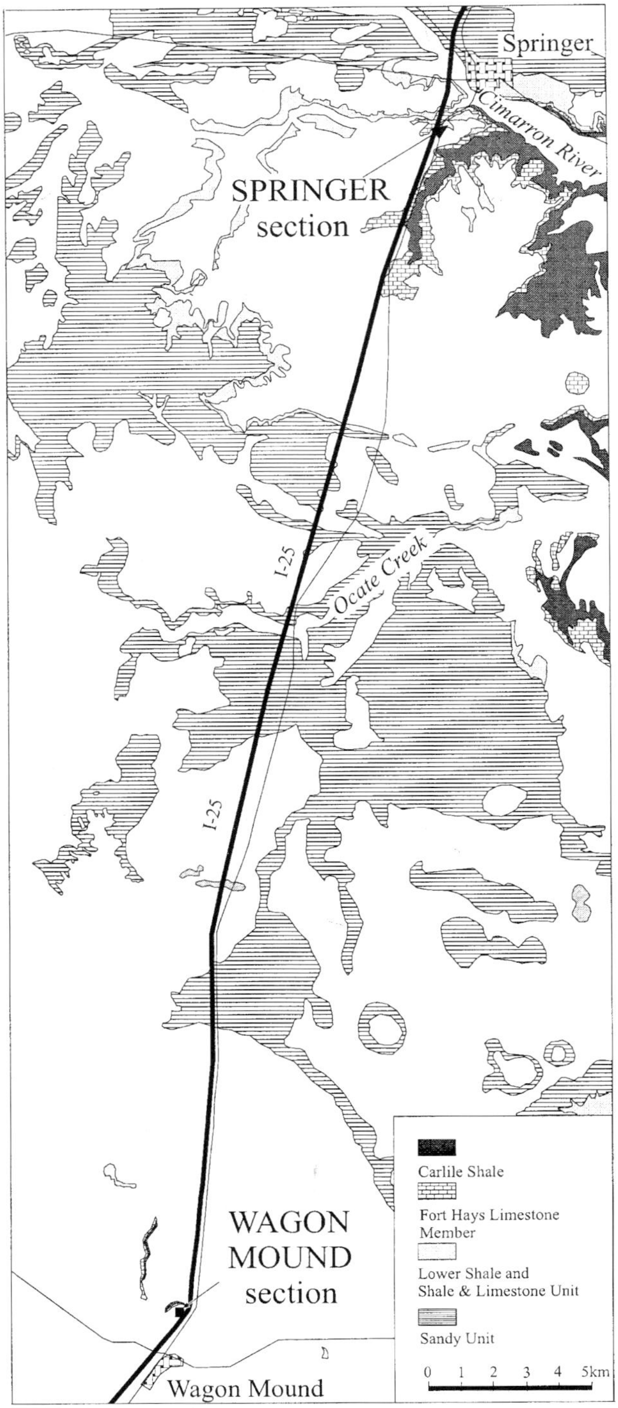

TEXT-FIG. 9. Geological sketch-map of the Springer-Wagon Mound area, north-eastern New Mexico; see Text-figure 1 for location in the general map of the Western Interior; modified from Scott (1986).

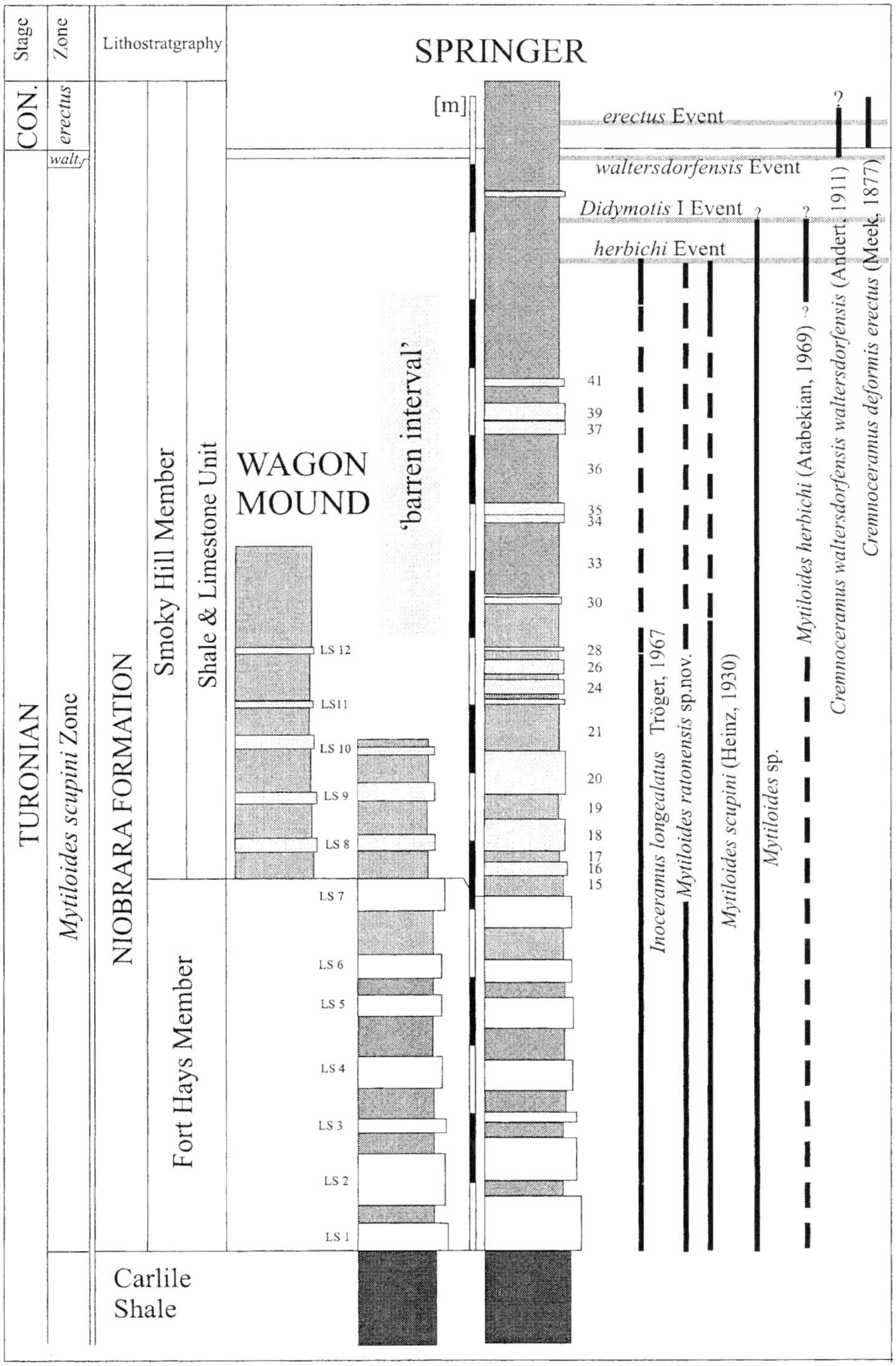

TEXT-FIG. 10. Geological columns, litho- and biostratigraphy, and the vertical distribution of inoceramids in the Springer and Wagon Mound sections; see Text-figure 9 for location. Numbers beside columns are bed numbers from measured sections.

Kennedy and Cobban 1991) is shown in Tables 1 and 2. The zonation is based on eight measured sections (Shelby, Johnson Ridge, North Belle Fourche, Spring Creek, Pueblo, La Junta, Springer and Wagon Mound), ranging geographically from northern Montana to northern New Mexico, discussed and illustrated in the previous section (Text-figs 3–10).

The correlation with the recently published inoceramid zonation of Kauffman *et al.* (1993; see also Kauffman *et al.* 1978, and in Larson *et al.* 1991) is presented in Table 1. This correlation is uncertain as many of Kauffman's index taxa are given in open nomenclature. As these were never described or illustrated, they are uninterpretable. Moreover, based on partly illustrated material (in Kauffman 1977*b*, and repeated in Kauffman *et al.* 1978), Kauffman's concepts of many species differ markedly from the original concepts. We have tried to compare our scheme as applied to the Pueblo section with the zonation applied to this section by Kauffman (after Kauffman and Pratt 1985). It appears that his concept of the Turonian/Coniacian boundary as well as the early stages of *Cremnoceramus* evolution are markedly different from those presented here (see discussion in Walaszczyk and Cobban 1999).

Inoceramus dimidius Interval Zone

Definition and character. The zone spans the interval from the FAD of the index taxon to the FAD of *Inoceramus perplexus* Whitfield, 1877. It is characterized by the monospecific occurrence of variable *I. dimidius* White, 1874. It is well developed in the North Belle Fourche section, and is represented by the thin Juana Lopez Member of the Carlile Shale in the La Junta section.

Correlation. The *dimidius* Zone corresponds to the *Prionocyclus macombi* and *Scaphites warreni* ammonite Zones. It corresponds to the *I. dimidius dimidius* and *I. dimidius* n. ssp. Zones (T11 and 12A' and 12A'' zones) of Kauffman *et al.* (1993).

Inoceramus perplexus Interval Zone

Definition and character. The lower boundary is defined by the FAD of *I. perplexus* Whitfield. The upper boundary is defined by the FAD of *Inocerams dakotensis* sp. nov.

The base of the *I. perplexus* Zone marks the main turnover in inoceramid fauna in the Middle/Upper Turonian boundary interval. The zone is characterized by the monospecific occurrence of the index species.

Correlation. The *I. perplexus* Zone corresponds to the lower and middle parts of the *Scaphites whitfieldi* Zone of Cobban (1951*b*).

In terms of Kauffman's zonation, the base of the zone corresponds to the base of his *I. perplexus perplexus* Zone (or the Zone 12A'''). The upper boundaries of the zones are not coeval. Although direct correlation is impossible (Kauffman never defined his *I. perplexus* ssp. nov.), the comparison is made through reference to the ammonite zonation. Kauffman (*in* Kauffman *et al.* 1978, 1993) equated the base of his *I. perplexus* n. ssp. Zone with the base of the *S. nigricollensis* Zone, which is actually higher than the top of the *I. perplexus* Zone as here defined.

Inoceramus dakotensis Interval Zone

Definition and character. The base of the zone corresponds to the FAD of *I. dakotensis* sp. nov.; the top is marked by the FAD of *Mytiloides incertus* (Jimbo, 1894). The zone is well represented in the North Belle Fourche section.

In contrast to the two previous zones, the *I. dakotensis* Zone is characterized by a variable inoceramid assemblage. Besides the index taxon, which probably evolved from *I. perplexus* (but is not its phyletic descendant as is indicated by the co-occurrence of both forms in this zone), *Mytiloides bellefourchensis* sp. nov. and *Mytiloides striatoconcentricus* (Gümbel, 1868) appear at the base of the zone. *Inoceramus perplexus* is still present in the lower part of the zone.

TABLE 1. Correlation of the inoceramid zonation determined in this paper with the standard ammonite scheme for the Western Interior, and inoceramid zonations as applied in Europe and the zonation applied to Western Interior by Kauffman (*in* Kauffman *et al.* 1993, and Kauffman and Pratt 1985).

STAGES	SUBSTAGES	AMMONOID ZONATION after Cobban 1951*b*, 1984*a* Kennedy and Cobban 1991 and this paper		US WESTERN INTERIOR INOCERAMID ZONATION after Kauffman *et al.* 1993 and Kauffman and Pratt 1985	INOCERAMID ZONATION this paper	EUROPEAN INOCERAMID ZONATION: after Ernst *et al.* 1983	EUROPEAN INOCERAMID ZONATION: sub-stages	EUROPEAN INOCERAMID ZONATION: Topmost Turonian - Lower Coniacian after Walaszczyk and Wood 1999*a*
CONIACIAN	Lower	*Forresteria alluaudi*	*Scaphites preventricosus*	*C. schloenbachi schloenbachi* *C. browni*, *C.* trans. *deformis* to *C. schloenbachi*	*Cremnoceramus crassus crassus*	*Cremnoceramus deformis*	LOWER CONIACIAN	*Cremnoceramus crassus crassus/C. deformis deformis*
		Forresteria hobsoni		*C. deformis* n.ssp. (late form)	*C. deformis dobrogensis* / *C. inconstans*	*Cremnoceramus erectus*		*Cremnoceramus c. inconstans* *Cremnoceramus walt. hannovrensis*
				C. deformis deformis *C. erectus* n.ssp. (late form) + *C.* n.sp. cf. *C. deformis*	*Cremnoceramus deformis erectus*	*Cremnoceramus rotundatus*		*Cremnoceramus deformis erectus*
TURONIAN	Upper		*Scaphites corvensis*	*C. erectus erectus*, *I. percostatus* *C? erectus* trans. *rotundatus*	*Cremnoceramus walt. waltersdorfensis*	*Cremnoceramus walt. hannovrensis*	UPPER TURONIAN	*Cremnoceramus walt. waltersdorfensis*
		Forresteria peruana		*C.? rotundatus*, *C.?*n.sp. cf. *C.? erectus*	*Mytiloides scupini*	*Inoceramus* aff. *frechi*		*Mytiloides scupini*
		P. germari		*Mytilodes dresdensis*, *M. lusatiae*	*Mytilodes incertus*	*Mytiloides labiatoidiformis* + *M. striatoconcentricus striatoconcentricus*		
		P. quadratus		*Mytilodes striatoconcentricus*				
		Scaphites nigricollensis		*Inoceramus perplexus* n.ssp.				
		Scaphites whitfieldi		*Inoceramus perplexus perplexus*	*Inoceramus dakotensis* *Inoceramus perplexus*	*I. costellatus* s.l. + *M. striat. striat.* *I.* ex gr. *cuvieri* + *I. c.* cf. *pietschi*		
		Scaphites warreni		*Inoceramus dimidius* n. ssp.	*Inoceramus dimidius*	*Inoceramus lamarcki* + *Inoceramus cuvieri*	MIDDLE TURONIAN	
		Prionocyclus macombi		*Inoceramus dimidius dimidius*				
	Middle	*Prionocyclus hyatti*		*Inoceramus howelli*, *Mytiloides costellatus*	*Inoceramus howelli*	?		

TABLE 2. Stratigraphical distribution of inoceramid faunas and selected ammonites in the Upper Turonian and Lower Coniacian of the Western Interior based on the sections studied.

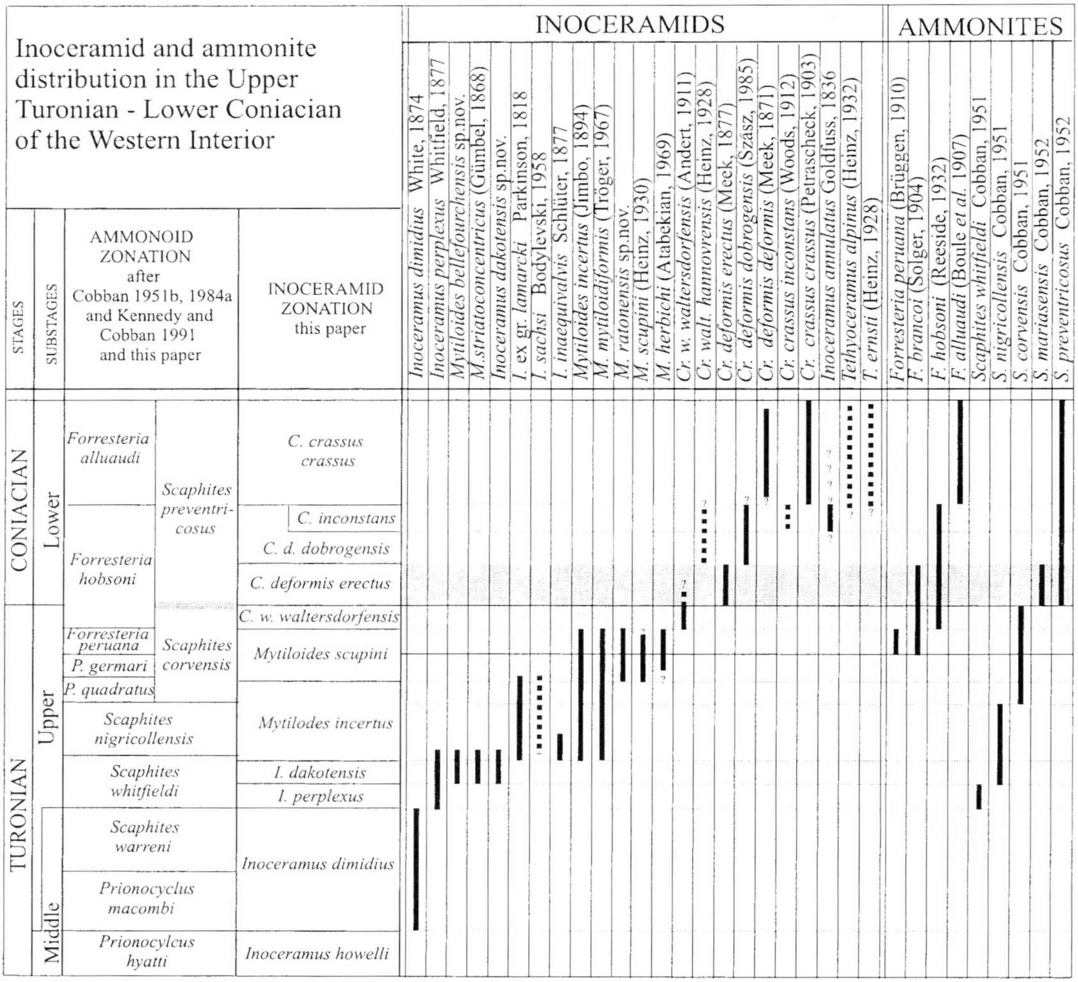

Correlation. The zone corresponds to the upper (but not uppermost) part of the *Scaphites whitfieldi* ammonite Zone. It corresponds to the upper part of the *I. perplexus perplexus* Zone of Kauffman (*in* Kauffman *et al*. 1978, 1993).

Mytiloides incertus Interval Zone

Definition and character. The base of the zone is defined by the FAD of the index taxon and its upper boundary by the FAD of *Mytiloides scupini* (Heinz, 1930). The index taxon is particularly common in the upper part of the zone, which corresponds to the base of the *Prionocyclus quadratus* ammonite Zone.

The zone is well documented in the North Belle Fourche and Spring Creek sections, South Dakota.

Many collections consisting almost entirely of *M. incertus* (Jimbo, 1984) come from the upper part of the zone, where there is an acme-occurrence of the species.

The lower part of the zone seems to be dominated by *M. mytiloidiformis* (Tröger, 1967) (see Cobban 1984*b*) and large representatives of the *lamarcki* group.

Correlation. The base of the zone lies in the topmost part of the *Scaphites whitfieldi* ammonite Zone. The upper part corresponds to the lower part of the *Prionocyclus quadratus* Zone (see Table 1; also Cobban 1984*b*, and Leckie *et al.* 1997).

The *M. incertus* Zone seems to be missing in Kauffman's scheme. The base of his *Mytiloides striatoconcentricus* Zone, according to Kauffman *et al.* (1993), is marked by the FAD of *M. aviculoides* and *M. kleini*? Both of these species are, however, misidentified. *M. aviculoides* is a lower Middle Turonian species (see Cobban and Reeside 1952*b*), and is one of the latest representatives of the Lower–lower Middle Turonian mytiloids with a widely, outwardly curved posterior auricle (atypical for the Early–early Middle Turonian mytiloids and similar to the Late Turonian representatives of the genus). *Inoceramus kleini* Müller, 1887, is an upper Lower–Middle Coniacian species and possesses very distinct characteristics, markedly different from forms referred to by Kauffman (1977*b*, and in Kauffman *et al.* 1978). Illustrated forms compared by Kauffman with Müller's species (Kauffman 1977*b*, pl. 9, fig. 24; pl. 10, figs 9, 15) should be referred to *M. ratonensis* sp. nov., or to some other latest Turonian mytiloids that are characteristic of the *M. scupini* Zone.

Mytiloides scupini Interval Zone

Definition and character. The base of the zone is defined by the FAD of *M. scupini* (Heinz, 1930); its top by the FAD of *Cremnoceramus waltersdorfensis waltersdorfensis* (Andert, 1911).

The lower part of the zone is characterized by a rich inoceramid fauna dominated by *M. scupini* (Heinz, 1930), *M. ratonensis* sp. nov., *M. herbichi* (Atabekian, 1969), *Mytiloides incertus* (Jimbo, 1894), and rare *Inoceramus longealatus* Tröger, 1967. This is succeeded by a barren interval, above which in the topmost part of the zone is a distinct level with *M. herbichi* and representatives of the bivalve genus *Didymotis*. This corresponds to the *Didymotis* I Event as recognized in Europe. *Inoceramus lusatiae* Andert, 1911, a very characteristic element of the Turonian/Coniacian boundary in Europe, was not found in the Western Interior fauna. Previous reports of this species (Kauffman 1977*b*, and *in* Kauffman *et al.* 1978, 1993; Collom 1991) are the result of different concepts of this species, and represent mostly *M. scupini* (Heinz, 1930).

Correlation. The zone corresponds to the *Prionocyclus germari* and *Forresteria peruana* ammonite Zones and to the lower and middle parts of the *Scaphites corvensis* ammonite Zone.

The correlation of this interval with Kauffman's scheme is very uncertain. His *M. striatoconcentricus* Zone corresponds most probably to the lower part of the *M. scupini* Zone. His succeeding zone of *M. dresdensis-M. lusatiae* contains inoceramids representing at least two stratigraphical intervals, the *M. scupini* Zone and the *C. waltersdorfensis* Zone, as used here. The base of his *M. dresdensis-M. lusatiae* Zone is defined by Kauffman *et al.* (1993) by the FAD of *C. waltersdorfensis waltersdorfensis* and *C. waltersdorfensis hannovrensis*. These two subspecies (treated by Kauffman apparently strictly typologically) do not appear, however, below the base of the *C. waltersdorfensis* Zone. At the same time, most of the species listed from that zone by Kauffman *et al.* (1993) characterize the *M. scupini* Zone or appear even well below this zone (as, for example, *M. incertus*). Based on the comparison of our scheme with the scheme that Kauffman (after Kauffman and Pratt 1985) applied to the Pueblo section (see Walaszczyk and Cobban 1999), the basal part of the *scupini* Zone as here defined corresponds to Kauffman's *Mytiloides dresdensis* + *M. lusatiae* Zone. The middle and upper parts of the *M. scupini* Zone correspond already to Kauffman's *C.* ? *rotundatus* + *C.*? n. sp. cf. *C.*? *erectus* Zone through *C. erectus erectus* + *I. percostatus* Zone, with the Turonian/Coniacian boundary indicated by the base of the *C.*? *rotundatus* + *C.*? n. sp. cf. *C.*? *erectus* Zone.

Cremnoceramus waltersdorfensis waltersdorfensis Interval Zone

Definition and character. This zone represents an interval from the FAD of the index taxon to the FAD of *C. deformis erectus* (Meek). The zone is characterized by an almost monospecific assemblage of the index taxon, and has been recognized in the Pueblo and La Junta sections.

Correlation. This zone corresponds to the topmost part of the *Scaphites corvensis* ammonite Zone and to the basal part of the *Forresteria hobsoni* ammonite Zone. In the Pueblo section Kauffman (*in* Kauffman and Pratt 1985), placed the interval here referred to the *waltersdorfensis* Zone to his zone of *I.* (*Cremnoceramus*?) *deformis* n. subsp. (Small early form) + *I. erectus*, n. ssp. (L. form) (see Walaszczyk and Cobban 1999).

Cremnoceramus deformis erectus Interval Zone

Definition and character. This zone represents an interval from the FAD of the index taxon to the FAD of *C. waltersdorfensis hannovrensis* (Heinz, 1928) or *C. deformis dobrogensis* (Szász, 1985).

The *erectus* Zone is characterized by the index taxon and, in the lower part, by rare representatives of *C. waltersdorfensis waltersdorfensis* (Andert, 1911). As in Europe, two distinct levels of flood occurrence of the index taxon are recognized; they correspond to the *erectus* I and *erectus* II events of European workers (see Walaszczyk and Wood 1999*a*). The zone is well represented in all of the sections studied.

Correlation. The *erectus* Zone is equivalent of the middle part of the *Forresteria hobsoni* ammonite Zone and the lower part of the *Scaphites preventricosus* ammonite Zone.

Based on a comparison with the scheme applied by Kauffman (*in* Kauffman and Pratt 1985) to the Pueblo section, the *erectus* Zone, as here defined, corresponds to his zone of *I.* (*Cremnoceramus*?) *deformis deformis* (see Walaszczyk and Cobban 1998).

Cremnoceramus deformis dobrogensis Interval Zone

Definition and character. The lower boundary of the zone is defined by the FAD of *C. deformis dobrogensis* (Szász, 1985) and its upper boundary by the FAD of *Cremnoceramus crassus crassus* (Petrascheck, 1903).

Inoceramids of the zone are dominated by the index taxon, which gives the impression of an almost monospecific assemblage. Rare *C. waltersdorfensis hannovrensis*, *C. crassus inconstans*, tethyoceramids, and *Inoceramus annulatus* were found in the Shelby section. A single specimen of *C. crassus inconstans* came from the upper part of the zone in the Pueblo section.

Correlation. The zone corresponds to the middle part of the *Scaphites preventricosus* ammonite Zone and upper part of the *Forresteria hobsoni* ammonite Zone. Comparing it with Kauffman's scheme through the Pueblo section (as published in Kauffman and Pratt 1985), the *dobrogensis* Zone corresponds to the upper part of Kauffman's *I.* (*Cremnoceramus*?) *deformis deformis* Zone and the lower part of his zone of *I.* (*Cremnoceramus*?) *deformis* n. ssp. (late form) (see Walaszczyk and Cobban 1999).

Cremnoceramus crassus inconstans Interval Zone

Definition and character. The base of the zone is defined by the FAD of the index taxon; its top marks the FAD of *C. crassus crassus* (Petrascheck, 1903).

Correlation. The zone corresponds to the topmost part of the *F. hobsoni* ammonite Zone and to the middle part of the *S. preventricosus* ammonite Zone. It is equivalent to the upper part of the *C. deformis dobrogensis* Zone.

Cremnoceramus crassus crassus Interval Zone

Definition and character. The base of the *crassus* Zone is defined by the FAD of the index species. Its upper boundary is marked by the FAD of volviceramids. It should be noted, however, that in the complete succession of Staffhorst, northern Germany, there is a distinct interval between the last cremnoceramids and the first volviceramids that is characterised by sulcate representatives of the *lamarcki* group (referred provisionally to *Inoceramus gibbosus* Schlüter, 1877) (Niebuhr *et al.* 1999). Whether or not a similar inoceramid succession occurs in the Western Interior requires further investigation.

Correlation. The *crassus* Zone corresponds to the upper part of the *S. preventricosus* ammonite Zone. The zone corresponds approximately to Kauffman's *C.* trans. *deformis* to *schloenbachi* Zone and *schloenbachi* Zone.

CORRELATION WITH EUROPE

The following correlation refers to the inoceramid zonation recently proposed by Walaszczyk and Wood (1999*a*) for the uppermost Turonian and Lower Coniacian, and by Ernst *et al.* (1983) for the Middle and lower Upper Turonian (see Table 1).

Other than some differences concerning the relative abundance of particular species, the interval from the base of the *M. scupini* Zone to the basal *crassus* Zone may be applied with small modifications to both Europe and the Western Interior. Further studies are needed, however, on the uppermost part of the Lower Coniacian corresponding to the upper *crassus* Zone and to the recently proposed *Inoceramus gibbosus* Zone (see Walaszczyk and Wood, *in* Niebuhr *et al.* 1999). This interval is not well represented in the studied material to allow a convincing comparison with the European succession

Below the *M. scupini* Zone (= *I. frechi* Zone of Ernst *et al.* 1983; see Walaszczyk and Tröger 1996), Ernst *et al.* (1983) distinguished three inoceramid zones. In ascending order these are the *I.* ex gr. *cuvieri*/*I. costellatus* cf. *pietschi* Zone, *I. costellatus* (*s.l.*)/*M. striatoconcentricus striatoconcentricus* Zone, and the *M. labiatoidiformis*/*M. striatoconcentricus striatoconcentricus* Zone. The base of the *I.* ex gr. *cuvieri*/*I. costellatus* cf. *pietschi* Zone, marked by the *costellatus*/*plana* Event, is placed at the FAD of the *costellatus* lineage, where it marks the base of the Upper Turonian as accepted in Europe. As discussed by Walaszczyk and Wood (1999*a*), forms referred to *I. costellatus* in Europe differ significantly from Woods' (1912) holotype, which is a *Mytiloides*. As shown below, the *I. costellatus* of authors *non* Woods, is conspecific with *Inoceramus perplexus* Whitfield, 1877, a species with a well-defined range interval in the Upper Turonian successions of the Western Interior. Accordingly, we correlate here the base of the *perplexus* Zone with the base of the *I. cuvieri*/*I. perplexus* (= *costellatus*) cf. *pietschi* Zone of Ernst *et al.* (1983).

Less convincing is the correlation of the base of the *I. dakotensis* Zone with the base of the *I. costellatus* (auctorum *non* Woods)/*M. striatoconcentricus striatoconcentricus* Zone. It is based on rare specimens of *M. striatoconcentricus* (Gümbel, 1868) and *Mytiloides bellefourchensis* sp. nov, found at the base of the *I. dakotensis* Zone in the North Belle Fourche section (see Text-fig. 6), species that also have their FAD at the base of the *I. costellatus* (auctorum *non* Woods)/*M. striatoconcentricus striatoconcentricus* Zone in Europe

In Europe, the record of *Mytiloides incertus* (Jimbo, 1894), is limited almost entirely to what is interpreted as a very narrow acme-interval, located within the *M. labiatoidiformis*/*M. striatoconcentricus striatoconcentricus* Zone (see Ernst *et al.* 1983). In the succeeding part of the Upper Turonian to the base of the *waltersdorfensis waltersdorfensis* Zone, *M. incertus* is known only from sporadic occurrences. In the Western Interior, the acme-occurrence of *M. incertus* falls relatively high within the range of this species (at the level corresponding to the base of the *Prionocyclus quadratus* ammonite Zone) and high within the *M. incertus* Zone as here defined. The base of the *M. incertus* Zone is provisionally correlated with the base of the *M. labiatoidiformis*/*M. striatoconcentricus striatoconcentricus* Zone, although the actual equivalence has not been demonstrated conclusively.

TEXT-FIG. 11. Map showing the localities of collections mentioned in the text.

SYSTEMATIC PALAEONTOLOGY

All specimens described in this paper are housed in the US National Museum of Natural History, Washington, DC, abbreviated USNM.

Generic classification

Late Turonian–early Coniacian inoceramids of the Western Interior represent at least four distinct genera: *Inoceramus* J. Sowerby, 1814, *Mytiloides* Brongniart, 1822, *Cremnoceramus* Cox, 1969 (*non* Heinz, 1932) and *Tethyoceramus* Sornay, 1980. These were discussed in detail in the recent literature (Kauffman *et al.* 1977; Kauffman, *in* Herm *et al.* 1979; Walaszczyk 1992; Crampton 1996; Harries *et al.* 1996; Walaszczyk and Wood 1999*a*), and although many questions are still open, they are currently used in a more or less consistent way. The most promising approach for understanding inoceramid evolutionary history is to study the sequence of taxa in the most complete inoceramid-rich strata. The study of interior structural characteristics of inoceramid shells, often quoted as the only basis for satisfactory classification, is illusory. Available data on the musculature is still insufficient for reliable use in classification, and the characteristics of the ligamentat and the ligament area may vary to such an extent (Crampton 1988, 1996; Walaszczyk 1997) that its simple application to inoceramid systematics appears unsound.

All forms with unknown generic affiliation are referred to the genus '*Inoceramus*' *sensu lato*.

Use of species and subspecies

Species rank is given to phylogenetic (cladogenetic) units, ranging between two speciation events, or between speciation event and final extinction. The co-occurrence of ancestral and daughter species is accepted. When phyletic (anagenetic) changes within a single species are well defined, successive morphotypes are referred to here as subspecies (=chronosubspecies), and the species is represented by the lineage. The name of the lineage is the oldest name available among the constituent chronosubspecies. Where a subspecies category is used in a neontological (geographical) sense, attention is drawn to the usage. Where relationships between close morphotypes are not clear, they are treated as separate species and their possible status is discussed.

Descriptive nomenclature

Terms commonly used to characterize the external morphology, as well as the basic structural elements of the inoceramid shell, are given in Text-figures 12–13; additional terms are defined in the text. Ornament elements are referred to as rugae and growth-lines. Where three-fold ornament is present, the term ribs is applied to elements subordinate to rugae. The sharp steps in the shell surface, occurring at the edges of rugae in cremnoceramids and tethyoceramids, are referred to growth marks (=*Anwachsmarken* of Heinz 1928*b*).

Family INOCERAMIDAE Giebel, 1852

Genus INOCERAMUS J. Sowerby, 1814

Type species. *Inoceramus cuvieri* J. Sowerby, 1814, by subsequent designation of Cox (1969, p. N315), from the Middle Chalk (*Terebratulina lata* Zone) of Royston, England.

Remarks. See Harries *et al.* (1996) for description and discussion of the genus.

Occurrence. ?Middle Cenomanian–basal Upper Coniacian world-wide.

Inoceramus dimidius White, 1874

Plates 1–4; Plate 8, figures 1, 4–6

1874 *Inoceramus dimidius* White, p. 25.
1877 *Inoceramus dimidius* White; White, p. 181, pl. 16, fig. 2a–d.
1893 *Inoceramus dimidius* White; Boyle, p. 155.
1894 *Inoceramus dimidius* White; Stanton, p. 78, pl. 10, figs 5–6.
1898 *Inoceramus dimidius* White; Logan, p. 452, pl. 98, figs 5–6.
1903 *Inoceramus dimidius* White; Johnson, p. 116, pl. 2, fig. 18.
1910 *Inoceramus dimidius* White; Grabau and Shimer, p. 441, fig. 579.
1942 *Inoceramus dimidius* White; Moreman, p. 200, pl. 31, figs 2, 6.
1944 *Inoceramus dimidius* White; Shimer and Schrock, p. 389, pl. 151, figs 3–4.
1975 *Inoceramus dimidius* White; Hattin, pl. 2, figs 6–7.
1977*b* *Inoceramus? dimidius dimidius* White; Kauffman, p. 238, pl. 8, figs 7, 12–13.
1978 *Inoceramus? dimidius dimidius* White; Kauffman *et al.*, pl. 12, figs 7, 12–13.
1980 *Inoceramus dimidius* White; Hook and Cobban, p. 44, fig. 7.
1986 *Inoceramus dimidius* White; Cobban, fig. 5C–D.
1989 *Inoceramus dimidius* White; Kennedy *et al.*, p. 186, fig. 33B–C, E, H–J.
1997 *Inoceramus dimidius* White; Leckie *et al.*, figs 33X–AA, 34K–M.
1997 *Inoceramus* n. sp. Leckie *et al.*, fig. 34B–J.

Type. The lectotype, by the subsequent designation of Kennedy *et al.* (1989, p. 106), is USNM 86232a, the original of White (1874, pl. 16, fig. 2b) from the Pescado Tongue of the Mancos Shale of the Pescado area, 21 km east of Zuni Pueblo, New Mexico (see Hook and Cobban 1980 for a review of the type locality problem). Three other species illustrated by White (1874, pl. 16, fig. 2a, c–d) are paralectotypes.

Material. The species is particularly well represented in the collection from locality 76 (D4395), having 68 double-valved adult specimens and 36 single-valved specimens (within the latter a half being represented by adults). In addition, 37 adult specimens come from locality 25 (21189) and about 80 juveniles from localities 24 (21188) and 23 (21187). The forms represented in most collections, as well as those usually found in the field, are small, often geniculated juveniles.

Dimensions (mm).	h	l	H	L	s	VR	α	δ	B	h max
USNM 501167	63	46·5	58	49·5	23	37	94	53	—	82·5
USNM 501171	61	49	57	53	28	34	90	56	—	84
USNM 501174	59·5	51·5	55·5	50	22	36	98	61	—	90
USNM 501202	71·5	60	66·5	56	25	47	95	54	—	100
USNM 501201	69	64	64·5	63	24	38·5	96	57	—	100
USNM 501213	74	66	69	66	42	29	96	50	—	114
USNM 501172	68·5	55	61	54	25	33	98	60	—	104
USNM 501170	55	41	46	48·5	23	27	97	53	—	69
USNM 501168	57	44·7	50·8	43	22	41	92	60	—	79

Emended diagnosis. Moderate- to large-sized, inequilateral, inequi- to semiequivalve, prosocline species with subpentagonal outline. Shell composed of juvenile and adult parts separated along negative geniculation. Juvenile part small, cup-shaped, usually well geniculated, variable. Adult part large, moderately inflated, with subrounded outline and often with indistinct sulcus in the posterior part of disc. Posterior auricle relatively small, in juveniles well separated from disc. Juvenile part ornamented with regular rugae; adult part smooth or with irregular rugae.

Description. Shells inequilateral, ranging from markedly inequivalve to semiequivalve. They consist of two different parts; a juvenile one, relatively small (forming approximately one-quarter of the adult size) and a large, flat adult (Pl. 1, figs 1, 4–7; Pl. 2, figs 2, 8–9; Pl. 3, figs 15–16; Pl. 4, figs 9–10). Juvenile part highly inflated, usually with well developed positive geniculation (Pl. 3, figs 9–11) (negative and positive geniculation are used here in the sense of

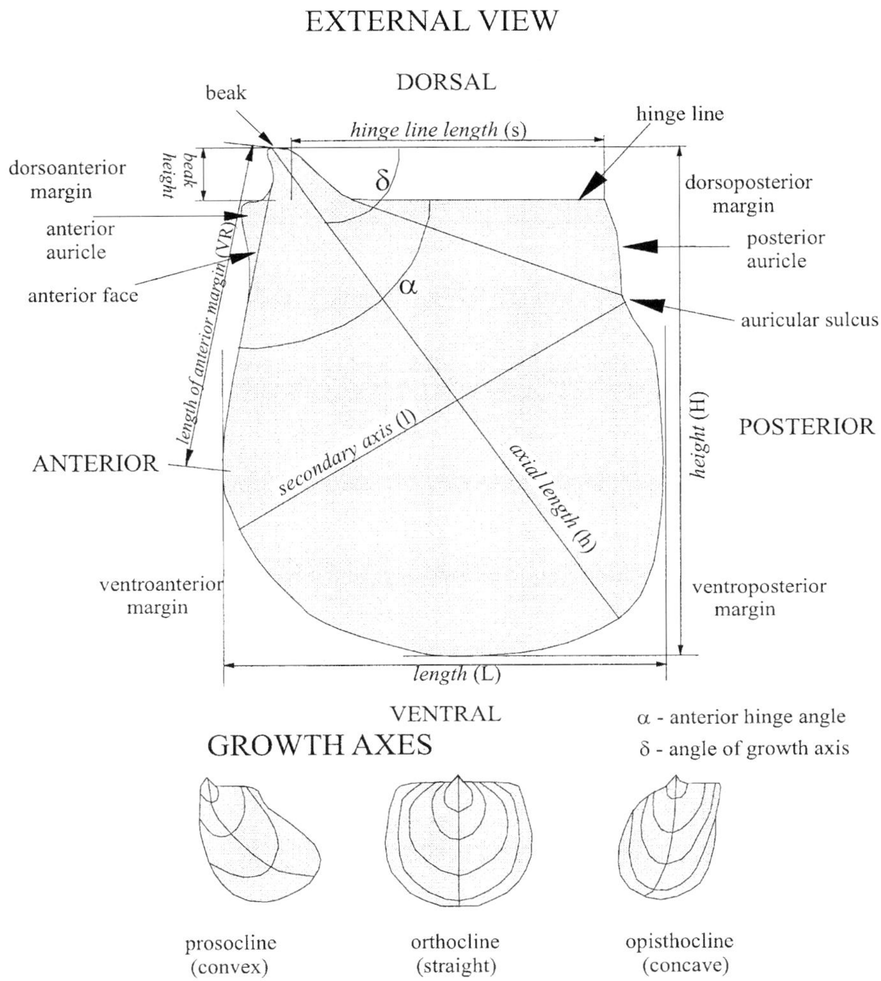

TEXT-FIG. 12. Terminology and measurements of the external morphologic features of inoceramid shells as applied in this paper (modified after Harries *et al.* 1996).

Crampton 1998; see also Text-figure 13 of this paper), dividing the juveniles into head and neck. The juvenile part is subrounded, markedly oblique, prosocline.

Umbonal parts of juveniles, or even whole juveniles, when relatively large, are ornamented with well-developed, regularly spaced rugae (e.g. Pl. 1, figs 5, 7; Pl. 2, figs 3, 5, 9; Pl. 3, figs 3, 7, 11–14, 17; Pl. 4, figs 4, 7, 11) and superimposed raised growth lines (see Pl. 2, figs 3, 5; Pl. 3, figs 3, 7, 11–14; Pl. 4, figs 4, 7, 11).

Adult part large with subpentagonal to triangular outline, moderately inflated, distinctly oblique. Anterior margin straight to slightly concave, long, usually passing into well-developed antero-ventral and ventral rounded margins. Posterior margin straight or slightly convex approaching hinge line at an angle 135–140 degrees. Hinge line straight, of moderate length or less (30 per cent of the relative axial length). In adult, posterior auricle not separated or only weakly separated from disc. Faint shallow sulcus present in axial part of disc in many specimens (Pl. 3, figs 15–16; Pl. 4, fig. 10). Adult ornamented with irregular, indistinct rugae (Pl. 1; Pl. 2, fig. 2) or almost smooth (Pl. 2, figs 7, 9).

Shelly specimens are extremely rare in the material studied. The ligament plate, observed only in a single specimen, has small, rectangular, shallow resilifers 1·8 mm high. Judging from the external moulds of the ligament plate observed in other specimens, this is a common type feature in the species.

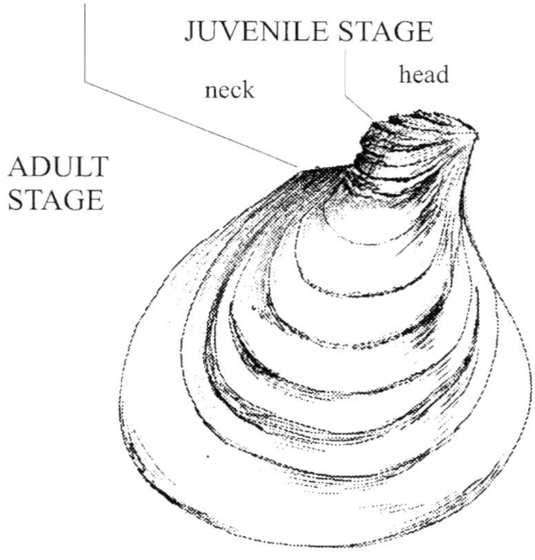

TEXT-FIG. 13. Terminology applied to geniculated form.

Remarks. *Inoceramus dimidius* is an evolutionary descendant of *Inoceramus howelli* White, 1876, and both forms are very similar. When compared to *I. howelli*, the adults of *I. dimidius* are much less inequivalve, are considerably broader, only rarely possess an indistinct axial sulcus, and adults are usually smooth, whereas *I. howelli* often retains the juvenile ornament throughout ontogeny. The inequivalve character of *I. howelli* is expressed by differences in the umbonal part, with the LV umbo being more inflated.

Inoceramus dimidius displays wide morphological variation, chiefly in the development of the juvenile part and the geniculation pattern. Specimens vary from massive, low forms with a very small juvenile stage and almost no neck, to slender, delicate morphotypes with a very high neck and poorly developed or no boundary between juvenile and adult. Forms that geniculate very early in ontogeny possess long and slender necks (Pl. 2, fig. 1; Pl. 4, figs 1, 3). When negative geniculation is poorly developed or not present, they closely resemble slender morphotypes of *I. howelli* or *Inoceramus inaequivalvis*. The latter, however, in contrast to White's species, has a massive ligamentat, and the umbonal part is not geniculated. The neck is sometimes very short, and the specimens have a dwarf appearance. Forms with a relatively large juvenile and a very short neck resemble *Inoceramus cuveri* (Pl. 1, fig. 4; Pl. 3, fig. 16; Pl. 4, figs 9–10). Extremely large juveniles may be easily misidentified with *Mytiloides incertus* (Jimbo, 1894) (see Pl. 2, figs 3, 5; Pl. 3, figs 13, 17; Pl. 4, fig. 4). The adults vary in h/l ratio and development of the sulcus in the posterior part of the disc corresponding to that which is usually well developed in *I. howelli* White.

Occurrence. *I. dimidius* occurs abundantly in the *Prionocyclus macombi* and *Scaphites warreni* zones in the Western Interior; not known outside North America.

EXPLANATION OF PLATE 1

Figs 1–7. *Inoceramus dimidius* White, 1874. 1, USNM 501169. 2, USNM 501183. 3, USNM 501184. 4, USNM 501170. 5, USNM 501171. 6, USNM 501168. 7, USNM 501172. 1, 4–7; USGS Mesozoic locality D4395 (Text-fig. 11, loc. 76). 2–3, USGS Mesozoic locality 21189 (Text-fig. 11, loc. 25). All ×1.

PLATE 1

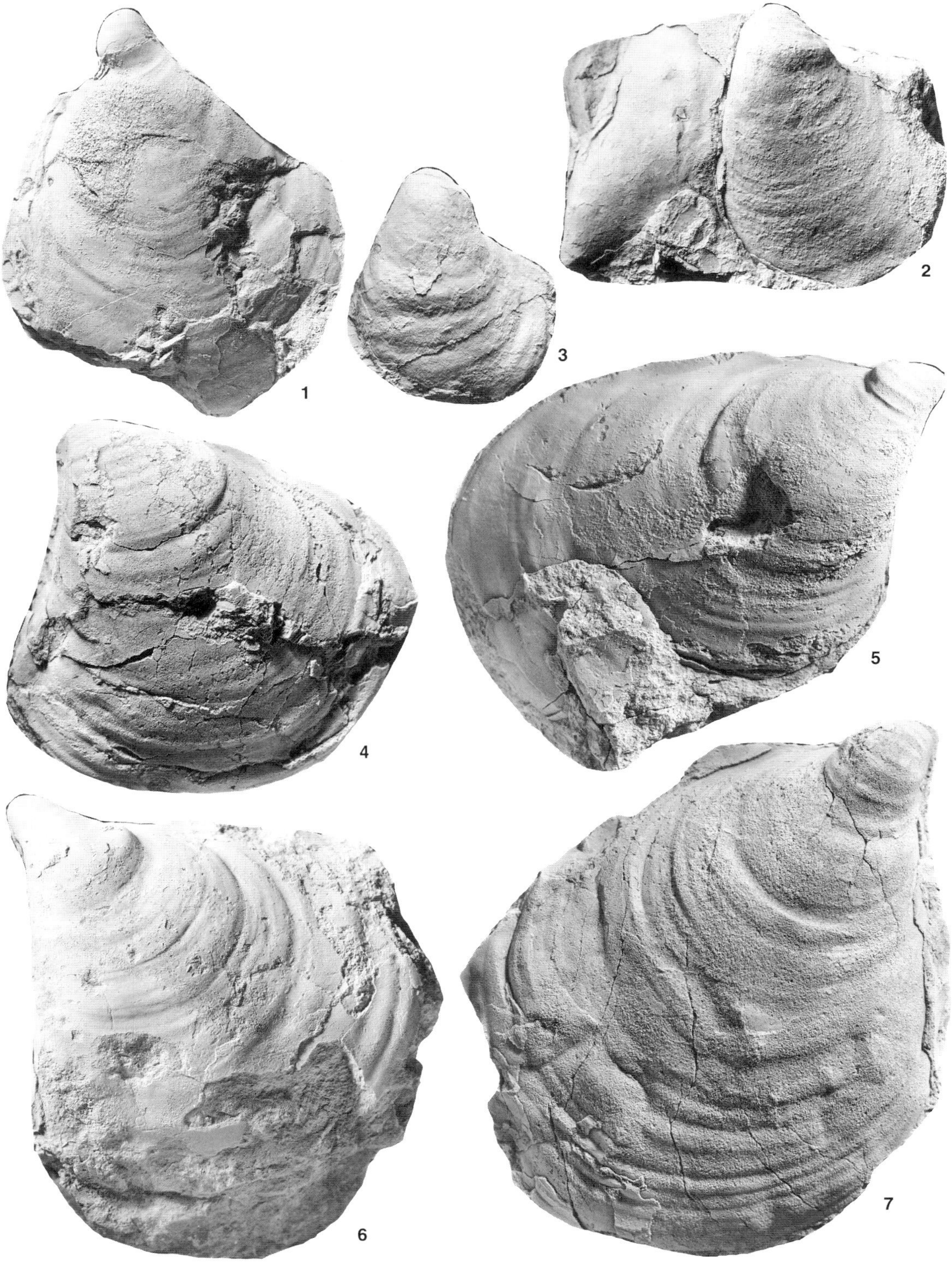

WALASZCZYK and COBBAN, *Inoceramus*

Inoceramus perplexus Whitfield, 1877

Plate 5, figures 1–9; Plate 6, figures 1–2; Plate 9, figure 7

1834-40	*Inoceramus undulatus* Mantell; A. Goldfuss, p. 115, pl. 112, fig. 1.
1877	*Inoceramus perplexus* Whitfield, p. 31.
1880	*Inoceramus perplexus* Whitfield; Whitfield, p. 392, pl. 8, fig. 3; pl. 10, figs 4–5.
1893	*Inoceramus perplexus* Whitfield; Boyle, p. 157.
1894	*Inoceramus fragilis* Hall and Meek; Stanton, p. 76 (*pars*), pl. 11, fig. 1.
1898	*Inoceramus perplexus* Whitfield; Logan, p. 450, pl. 87, fig. 1.
1912	*Inoceramus costellatus* Woods, p. 336 (*pars*), pl. 54, fig. 7 only, *non* pl. 54, figs 5–6 (= *M. costellatus*).
1930	*Inoceramus costellatus* Woods; Fiege, p. 35, pl. 5, figs 3–7, 9.
?1964	*Inoceramus* aff. *I. perplexus* Whitfield; Scott and Cobban (*pars*), pl. 2, fig. 5.
1967	*Inoceramus vancouverensis vancouverensis* Shumard; Tröger, p. 89, pl. 9, figs 6–9.
1967	*Inoceramus vancouverensis parvus* Tröger, p. 92, pl. 9, figs 1–5, pl. 10, fig. 3.
1977*b*	*Inoceramus* (*Inoceramus*) *perplexus* Whitfield; Kauffman, p. 238, pl. 8, figs 6, 15.
1978	*Inoceramus* (*Inoceramus*) *perplexus* Whitfield; Kauffman *et al.*, pl. 12, figs 6, 15.
1978*b*	*Inoceramus* (*Inoceramus*) n. sp. aff. '*I. costellatus* Woods' of Fiege, 1930, pl. 5, fig. 10, and *I. uwajimensis* Yehara, 1924, pl. 3, fig. 2; fig. 4, fig. 2; Kauffman, pl. 2, figs 1, 4, 8.
1984*b*	*Inoceramus perplexus* Whitfield; Cobban, p. 11, pl. 2, figs 1–3.
1982	*Inoceramus costellatus costellatus* Woods; Keller, p. 92, pl. 7, fig. 2.
1992	*Inoceramus costellatus* Woods, 1911; Walaszczyk, p. 31, pl. 12, figs 3–9.

Types. The lectotype, by subsequent designation of Kennedy *et al.* (1989), is USNM 12274, the original of Whitfield (1880, pl. 8, fig. 3 and pl. 10, fig. 5), re-illustrated here in Plate 6, figure 2, from the Carlile Shale. According to Whitfield, it came from 'Belle Fourche River, 10 miles west of Crow Peak, Black Hills'. Ten miles north of Crow Peak is, however, in Carboniferous rocks (De Witt *et al.* 1989). More likely the specimens are from the Belle Fourche area of South Dakota 17 or 18 miles (27–28 km) north of Crow Peak. The second specimen of Whitfield (1880, pl. 10, fig. 4), re-illustrated here in Plate 6, figure 1 (USNM 12263), is from the same locality as the lectotype.

Material. Numerous specimens from locality 28 (21294), 29 (21295), 57 (D6928), 55 (D9010), 84 (D7002), and many compressed specimens from locality 73 (D9232). Plaster casts of the lectotype and paralectotype.

Dimensions (*mm*).	h	l	H	L	s	VR	α	δ	B	h max
Lectotype	37	31	34·7	28	20·5	25·5	98	70	—	41·4
Paralectotype	38·7	33·4	—	—	—	25	107	72	—	41·3
USNM 501219	41·1	37·1	40	38	21	24	108	68	—	54
USNM 501228	37·2	30	34.6	29·5	19·5	22·5	105	79	—	54
USNM 501230	55	44	50	42	26·5	34·5	105	70	—	68
USNM 501231	34	25	31·8	25·5	18·5	21	100	65	—	36·5
USNM 501216	27·8	24	26	21·3	17·1	16·2	100	74	—	30
USNM 501229	36·6	30	34	28·6	24	26	99	72	—	42
USNM 501218	30·1	21·2	28	26·1	15·5	21·4	101	65	—	37

Diagnosis. Small to moderate-sized, orthocline, with anterior margin almost straight, slightly concave near the umbo, and rounded ventral and posterior margins. Posterior auricle not separated from disc, small.

EXPLANATION OF PLATE 2

Figs 1–9. *Inoceramus dimidius* White, 1874. 1, USNM 501175, USGS Mesozoic locality D9244 (Text-fig. 11, loc. 58). 2, USNM 501173, USGS Mesozoic locality D4395 (Text-fig. 11, loc. 76). 3, USNM 501176, USGS Mesozoic locality 21187 (Text-fig. 11, loc. 23). 4, USNM 501180, USGS Mesozoic locality 21189 (Text-fig. 11, loc. 25). 5, USNM 501179, USGS Mesozoic locality 21187 (Text-fig. 11, loc. 23). 6, USNM 501178, USGS Mesozoic locality D3763 (Text-fig. 11, loc. 56). 7, USNM 501181, USGS Mesozoic locality 21189 (Text-fig. 11, loc. 25). 8, USNM 501167; 9, USNM 501174: USGS Mesozoic locality D4395 (Text-fig. 11, loc. 76). All ×1.

PLATE 2

WALASZCZYK and COBBAN, *Inoceramus*

Valves covered with regularly, evenly spaced concentric rugae, with gradual increase of inter-rugae spaces ventral ward. Growth lines not visible.

Description. The species is usually represented by small specimens (h max 30–40 mm). Much rarer specimens attain 6–7 or even 10 cm in h max.

Outline subquadrate, equivalve, inequilateral, weakly to moderately inflated with maximum inflation in dorsal part. Anterior margin long, straight, slightly concave at umbo, passing into broadly rounded ventral margin. Anterior face steep, perpendicular to commissural plane. Hinge line straight, moderately long to long (S/h ratio ranges from 0·50–0·66). Umbonal region raised, beak curved dorso-anteriorly, only slightly projecting above hinge line. Posterior auricle small, usually not separated from disc.

The species is ornamented with regular, evenly spaced concentric rugae, with inter-rugae spaces increasing gradually ventrally. Rugae sharp-edged, narrow. Umbonal part up to 15 mm axial length, covered exclusively with raised growth lines.

Shelly material very rare, with the exception of collection 10905 in which many specimens are preserved with inner nacreous layer. Ligament plate rarely preserved, shows a very delicate, 2-mm-high plate with evenly spaced, rectangular resilifers that have very minute inter-resilifer spaces.

Remarks. As well shown by the types (Pl. 6, figs 1–2), *Inoceramus perplexus* is characterized by a very regular shape and regular ornament composed of concentric rugae that gradually increase in size and spacing through ontogeny. Growth lines are usually absent. Both of these characters are retained even in the largest specimens. Much less regular forms with quite irregular ornament and outline were formerly referred to *I. perplexus*. These are now assigned to *Inoceramus dakotensis* sp. nov. (see below).

Inoceramus perplexus is the senior synonym of European forms usually referred to *Inoceramus costellatus*, but the holotype of Woods' species is a *Mytiloides* (see discussion in Walaszczyk and Wood 1999*a*).

Whitfield's species resembles the Japanese species *Inoceramus teshioensis* Nagao and Matsumoto (1939, p. 274, pl. 24, figs 6–7, 9, and pl. 26, figs 5–7) and *Inoceramus tenuistriatus* Nagao and Matsumoto (1939, p. 272, pl. 24, fig. 8 and pl. 26, figs 1–4) in valve outline. It differs from both in the ornament, which in Whitfield's species is composed of concentric rugae only, whereas *I. teshioensis* has ribs between the main rugae (see Nagao and Matsumoto 1939, pl. 24, figs 6, 9). Some rugate specimens of *I. tenuistriatus* (Nagao and Matsumoto 1939, pl. 26, fig. 4) may be very similar to specimens of *I. perplexus* with less pronounced rugae.

Based on the European record, Tröger (1967, 1981*b*, 1989) distinguished within *I. perplexus* (=*I. costellatus* of authors) two chronosubspecies, *I. costellatus costellatus* and *I. costellatus pietschi* (which should be referred to *Inoceramus perplexus perplexus* and *I. perplexus pietschi*). These two morphotypes are well represented in the material studied, but seem to co-occur throughout the whole of the stratigraphical range of Whitfield's species, of which they are regarded as synonyms.

The appearance of *I. perplexus* marks the main change in the inoceramid fauna in the Middle/Upper Turonian boundary interval in the Western Interior as well as in the whole Euramerican biogeographical region. In the Western Interior the entrance level of *I. perplexus* coincides with the upper occurrence limit of the *I. howelli* – *I. dimidius* lineage, and it is succeeded by the appearance of the Late Turonian *Mytiloides* fauna.

EXPLANATION OF PLATE 3

Figs 1–18. *Inoceramus dimidius* White, 1874. 1, USNM 501182, USGS Mesozoic locality 21189 (Text-fig. 11, loc. 25). 2, USNM 501185, USGS Mesozoic locality 21187 (Text-fig. 11, loc. 23). 3, USNM 501187, USGS Mesozoic locality 21189 (Text-fig. 11, loc. 25). 4, USNM 501177, USGS Mesozoic locality 21187 (Text-fig. 11, loc. 23). 5, USNM 501190, USGS Mesozoic locality D3763 (Text-fig. 11, loc. 56). 6, USNM 501186, USGS Mesozoic locality D3763 (Text-fig. 11, loc. 56). 7, USNM 501188, USGS Mesozoic locality 21189 (Text-fig. 11, loc. 25). 8, USNM 501189, USGS Mesozoic locality 21187 (Text-fig. 11, loc. 23). 9, USNM 501191; 10, USNM 501192; 11, USNM 501193: all USGS Mesozoic locality 21187 (Text-fig. 11, loc. 23). 12, USNM 501196, USGS Mesozoic locality 21189 (Text-fig. 11, loc. 25). 13, USNM 501197, USGS Mesozoic locality D4395 (Text-fig. 11, loc. 76). 14, USNM 501195, USGS Mesozoic locality 21189 (Text-fig. 11, loc. 25). 15, USNM 501202; 16, USNM 501201: USGS Mesozoic locality D4395 (Text-fig. 11, loc. 76). 17, USNM 501198, USGS Mesozoic locality D4395 (Text-fig. 11, loc. 76). 18, USNM 501194, USGS Mesozoic locality D3763 (Text-fig. 11, loc. 56). All ×1.

WALASZCZYK and COBBAN, *Inoceramus*

Occurrence. Known from the Upper Turonian of the Euramerican biogeographical region. In the Western Interior it has its FAD at the base of the *Scaphites whitfieldi* Zone and ranges into the *S. nigricollensis* Zone.

Inoceramus dakotensis sp. nov.

Plate 5, figure 11; Plate 6, figures 3, 5–7, 11–15; Plate 7, figures 4–5; Plate 9, figure 4

1977*b* *Inoceramus (Inoceramus) perplexus* n. subsp. (late form); Kauffman, p. 238, pl. 8, fig. 14.
1978 *Inoceramus (Inoceramus) perplexus* n. subsp. (late form); Kauffman *et al*., pl. 12, fig. 14.
1984*b* *Inoceramus perplexus* Whitfield; Cobban, p. 11, pl. 2, figs 1–3.
?1997 *Inoceramus perplexus* Whitfield; Leckie *et al*., fig. 34O–R.
1997 *Inoceramus longealatus* Tröger; Leckie *et al*., fig. 36E–G.

Derivation of name. From South Dakota.

Types. The holotype is USNM 501227 (Pl. 5, fig. 11) from locality 29 (21195). Paratypes are USNM 501235–501239, 501243–501245, 501251, and 501252, all from USGS Mesozoic locality 28 (21194); USNM 50246 from USGS Mesozoic locality 29 (21195); and USNM 501266, 501290, 501271, and 501250 from USGS Mesozoic locality 84 (D7002).

Type locality. North Belle Fourche section (locality 29, USGS Mesozoic locality 21195), north-east margin of the Black Hills, South Dakota.

Type horizon. Turner Member of the Carlile Shale, *Scaphites whitfieldi* ammonite Zone.

Material. Numerous specimens from locality 28 (21194), 29 (21195), 84 (D7002), 72 (18703), 68 (D7252), and 64 (D3024).

Dimensions (*mm*).	h	l	H	L	s	VR	α	δ	B	h max
USNM 501243	52	39·5	51	37	20·5	40·3	104	78	12	72
USNM 501235	30	24·2	—	—	10	21·5	118	102	10	31·6
USNM 501237	35·4	28	—	—	—	28	—	—	—	37
USNM 501236	36·7	27	—	—	—	27	—	—	—	38
USNM 501244	55	40·8	—	—	—	39·3	—	—	—	64
USNM 501238	49	37	43	33	—	31·4	90·2	65	13	49
USNM 501251	64·6	46·5	—	—	28	49·4	110	76	—	73·3
USNM 501252	50	43·6	48·2	41·6	29	35	111	69	18	64
USNM 501247	55	44·5	51	42	27	38·8	100	65	11·5	65
USNM 501271	88	66	87	59·4	43	62	103	69	19·8	98
USNM 501250	87·8	61·7	82	59	45	67	92	62	21	116

EXPLANATION OF PLATE 4

Figs 1–13. *Inoceramus dimidius* White, 1874. 1, USNM 501203, USGS Mesozoic locality 21189 (Text-fig. 11, loc. 25). 2, USNM 501214, USGS Mesozoic locality D3763 (Text-fig. 11, loc. 56). 3, USNM 501205, USGS Mesozoic locality 21189 (Text-fig. 11, loc. 25). 4, USNM 501208, USGS Mesozoic locality 21187 (Text-fig. 11, loc. 23). 5, USNM 501200, USGS Mesozoic locality D3763 (Text-fig. 11, loc. 56). 6, USNM 501212, USGS Mesozoic locality D3763 (Text-fig. 11, loc. 56). 7, USNM 501204, USGS Mesozoic locality 21187 (Text-fig. 11, loc. 23). 8, USNM 501210, USGS Mesozoic locality 21189 (Text-fig. 11, loc. 25). 9, USNM 501215; 10, USNM 501213; USGS Mesozoic locality D4395 (Text-fig. 11, loc. 76). 11, USNM 501209, USGS Mesozoic locality 21187 (Text-fig. 11, loc. 23). All × 1.

WALASZCZYK and COBBAN, *Inoceramus*

Diagnosis. Upright, axially elongated, small to medium-sized species, with extended posterior auricle well separated from disc. Beak projected above hinge line, pointed. Small forms usually geniculated, with small juvenile part and long adult part. Anterior margin large, straight to slightly concave. Ornament composed of evenly and more or less closely spaced concentric ribs with superimposed wide and low concentric rugae.

Description. Small to moderate size, slightly inequivalve, inequilateral. Valve outline subquadrate to subpentagonal, axially elongated. Growth axis straight. Both valves differ slightly in juvenile obliquity with right valve being more oblique. Different obliquity causes differences in development of posterior auricle; auricular sulcus more pronounced in the more oblique right valve. Beak pointed and projected above hinge line. Juvenile stage 15–20 mm long, separated from rest of shell by a well-developed geniculation. Anterior margin straight or slightly concave, very long, up to 75–80 per cent of corresponding axial length. Anterior face steep. Ventral margin rounded, passing into wide postero-ventral margin and then into posterior margin. Hinge line straight, relatively long. Posterior auricle large, well separated from disc by well-developed auricular sulcus.

Shell ornamented by more or less evenly and closely spaced concentric ribs with superimposed wide, low, and less regular concentric rugae. As concentric elements pass onto posterior auricle they curve posteriorly and form a spindle-shaped lobe at postero-dorsal margin of auricle.

Remarks. The species displays moderate variability in ornament (regularity and density of concentric elements), l/h ratio (although this may also be partly a result of preservation), presence/absence of geniculation, and size of the posterior auricle. The latter two characters are interrelated; as a result of shell geometry geniculated forms usually have a larger posterior auricle, well separated from the disc. USNM 501266 (Pl. 9, fig. 4) has an exceptionally large posterior auricle; it closely resembles *Inoceramus longealatus* Tröger, 1967 in shape and overall ornament. Tröger's species may well be an evolutionary descendant of *Inoceramus dakotensis* sp. nov.

Small forms are very often geniculated. The geniculation is probably more common in *I. dakotensis* sp. nov. than indicated by the material studied. All three-dimensionally preserved specimens are geniculate, and its absence in some specimens may be the result of post-mortem crushing.

Ornament ranges from fine to coarse ribbing (compare Pl. 6, figs 5–7, 11–15 with Pl. 7, figs 4–5, and Pl. 9, fig. 4). Irregular, low rugae are usually present in adults. The phyletically oldest specimens are small with weak or no rugae (Pl. 6, figs 5–7).

A series of finely ornamented specimens from localities 68 (D7252) and 72 (18703) (Pl. 7, fig. 3; Pl. 11, figs 1–2, 4) match *I. dakotensis* sp. nov. in all other respects and are here referred to as *Inoceramus* aff. *dakotensis* sp. nov.

Inoceramus dakotensis sp. nov. is an evolutionary descendant of *Inoceramus perplexus* Whitfield (see above). As both forms overlap in part of their ranges (see Text-fig. 6), they should be regarded as two separate species rather than two chronosubspecies.

Judging from Kauffman's illustration of *I. perplexus* n. subsp. (late form) (*in* Kauffman *et al.* 1978, pl. 12, fig. 14), his form is conspecific with *I. dakotensis* sp. nov. Specimens of *I. dakotensis* have usually

---

EXPLANATION OF PLATE 5

Figs 1–9. *Inoceramus perplexus* Whitfield, 1877. 1, USNM 501218, USGS Mesozoic locality D6928 (Text-fig. 11, loc. 57). 2, USNM 501219, USGS Mesozoic locality 21194 (Text-fig. 11, loc. 28). 3, USNM 501220, USGS Mesozoic locality D9010 (Text-fig. 11, loc. 55). 4, USNM 501216, USGS Mesozoic locality D6928 (Text-fig. 11, loc. 57). 5, USNM 501222, USGS Mesozoic locality D6928 (Text-fig. 11, loc. 57). 6, USNM 501217, USGS Mesozoic locality D6928 (Text-fig. 11, loc. 57). 7, USNM 501224; 8, USNM 501225: both USGS Mesozoic locality D9232 (Text-fig. 11, loc. 73). 9, USNM 501223, USGS Mesozoic locality D6928 (Text-fig. 11, loc. 57).

Fig. 10. *Inoceramus inaequivalvis* Schlüter, 1877, USNM 501226, USGS Mesozoic locality 21195 (Text-fig. 11, loc. 29); a, anterior view; b, lateral view.

Fig. 11. *Inoceramus dakotensis* sp. nov., USNM 501227, holotype, USGS Mesozoic locality 21195 (Text-fig. 11, loc. 29).

All $\times 1$.

WALASZCZYK and COBBAN, *Inoceramus*

been referred to *Inoceramus perplexus* (see Cobban 1984*b*, pl. 2, figs 1–3; Leckie *et al.* 1997, fig. 34O–R); small forms were referred to *Inoceramus longealatus* Tröger by Leckie *et al.* (1997, fig. 36E–G).

Occurrence. In the North Belle Fourche section, the type locality, *I. dakotensis* sp. nov. appears at locality 26 (D21190), low within the *Scaphites whitfieldi* Zone, and ranges to locality 29 (D21195), close to the boundary with the *Scaphites nigricollenis* ammonite Zone; so far not known outside the Western Interior.

Inoceramus lamarcki Parkinson, 1818 group

Remarks. The term *lamarcki* group encompasses forms corresponding to Woods' (1912) broad concept of *I. lamarcki* Parkinson, 1818, a group of specimens with poorly understood evolutionary relationships. We also include in the *lamarcki* group the inequivalve forms (e.g. *I. inaequivalvis* Schlüter) as well as radially sulcate forms. The sulcate representatives (or those with a clear tendency to develop a radial sulcus) appeared at least twice in the history of the group. In the Middle and early Late Turonian, sulcate forms were common in the western North Pacific Province (forms referred to *I. hobetsensis* Nagao and Matsumoto, 1939, p. 281, and *I. iburiensis* Nagao and Matsumoto, 1939, p. 291) and in the North American Western Interior (forms referred to *I. flaccidus* White, 1876, p. 178, and *I. howelli* White, 1876, p. 114). Many European species, including the holotype of *I. lamarcki* Parkinson (Parkinson 1818, pl. 1, fig. 3; see also Woods 1912, text-fig. 63), also show a clear tendency to develop a radial sulcus. In the latest Early and Middle Coniacian, sulcate *lamarcki* forms were common in Europe and are referred variably to *I. percostatus* Müller, 1888, *I. gibbosus* Schlüter, 1877, *I. lezennensis* Decocq, 1874, and *I. russiensis* Nikitin, 1888. Both these and the North American forms are in need of thorough revision (see also discussion by Walaszczyk and Wood, *in* Niebuhr *et al.* 1999).

The group occurs over all of the Northern Hemisphere (Pergament 1978; Matsumoto and Noda, *in* Noda and Muramoto 1980) and displays great morphological plasticity. About 50 species and subspecies have been formally described within the group, and although many of them appear to be synonymous, the large number of named taxa illustrates the high variability, the nature of which needs further study. The group first appears definitely at the base of the Middle Turonian (possibly even lower) and ranges high into the Coniacian (well-dated Coniacian representatives of the group range to the beginning of the Late Coniacian; Walaszczyk and Wood, *in* Niebuhr *et al.* 1999).

There are two different opinions concerning the biogeography of the group. The most common view, expressed explicitly by Matsumoto (1981) and Matsumoto and Noda (*in* Noda and Muramoto 1980), assumes that, in spite of marked similarities, the group is represented by different species assemblages in different faunal provinces in the Northern Hemisphere. The other opinion, advocated mostly by Kauffman (1977*a*, 1978*c*; and Kauffman *et al.* 1978) focuses on the high morphological similarity and presumed taxonomical identity of *lamarcki* faunas within the entire North Temperate Realm. Illustrated forms of the

---

EXPLANATION OF PLATE 6

Figs 1–2. *Inoceramus perplexus* Whitfield, 1877. 1, USNM 12263, plaster cast of the paralectotype, original of Whitfield (1880, pl. 10, fig. 4). 2, USNM 12274, plaster cast of the lectotype, original of Whitfield (1880, pl. 8, fig. 3; pl. 10, fig. 5)

Figs 3, 5–7, 11–15. *Inoceramus dakotensis* sp. nov. 3, USNM 501237; 5, USNM 501235; 6, USNM 501236; 7, USNM 501239; 11, USNM 501245; 12, USNM 501243: all USGS Mesozoic locality 21194 (Text-fig. 11, loc. 28). 13, USNM 501246, USGS Mesozoic locality 21195 (Text-fig. 11, loc. 29). 14, USNM 501238; 15, USNM 501244: USGS Mesozoic locality 21194 (Text-fig. 11, loc. 28).

Figs 4, 9–10. *Mytiloides bellefourchensis* sp. nov. 4, USNM 501241; 9, USNM 501240; 10, USNM 501242: all USGS Mesozoic locality 21195 (Text-fig. 11, loc. 29).

Fig. 8. *Inoceramus* aff. *dakotensis* sp. nov., USNM 501247, USGS Mesozoic locality 21195 (Text-fig. 11, loc. 29). All × 1

PLATE 6

WALASZCZYK and COBBAN, *Inoceramus, Mytiloides*

same morphotypes, occupying probably the same stratigraphical position, seem to be present over all of the Northern Hemisphere. The existing data on the group are, however, so far insufficient to judge whether we really have widely distributed species or that the same morphotypes are repeated in the group in a mosaic way in various parts of the record of the group in both space and time.

In Europe, the group appears near the base of the Middle Turonian *Collignoniceras woollgari* ammonite Zone, and definitely ranges to the base of the Upper Coniacian. Its later history is unclear but it has not been reported from stratigraphically higher levels in the Euramerican biogeographic region at least. It is very well represented in the Old World, where it dominates the Middle Turonian record. It is still common, however, in the Upper Turonian as well as in the middle and upper Lower Coniacian.

In the material studied from the US Western Interior, representatives of the group were found in the Upper Turonian, in the basal *M. incertus* Zone (base of the *Scaphites nigricollensis* ammonite Zone), and single specimens were found higher in the middle Lower Coniacian of the Shelby section. The Upper Turonian forms are represented by a relatively variable assemblage (*I. inaequivalvis* Schlüter, 1877, *I.* cf. *stuemckei* Heinz, 1928, *I.* cf. *cuvieriformis* Pergament, 1971) (Text-fig. 14; Pl. 5, fig. 10; Pl. 8, figs 2–3, 7; Pl. 9, figs 5–6; Pl. 10, figs 6–7). The Coniacian specimens correspond very well to the European representatives of large, upright forms (Pl. 31, fig. 3), referred to *I. annulatus* Goldfuss, 1836 (see Walaszczyk and Wood 1999a).

Inoceramus inaequivalvis Schlüter, 1877

Plate 5, figure 10

1834–40	*Inoceramus striatus* Mantell; Goldfuss, pl. 112 (*pars*), fig. 2d–e.
1877	*Inoceramus inaequivalvis* Schlüter, p. 265.
1912	*Inoceramus lamarcki* Parkinson; Woods, p. 311 (*pars*), pl. 52, figs 4–6.
1928*a*	*Inoceramus inaequivalvis* Schlüt. var. *falcata* Heinz, p. 72.
1932*a*	*Inaequiceramus modestus* Heinz, p. 35.
?1959	*Inoceramus seitzi* Andert; Dobrov and Pavlova, p. 143, pl. 3, fig. 3.
1967	*Inoceramus inaequivalvis inaequivalvis* Schlüter, 1877; Tröger, p. 79, pl. 7, figs 1–2, 6.
1967	*Inoceramus inaequivalvis modestus* Heinz; Tröger, p. 82, pl. 7, figs 7–8.
1968	*Inoceramus falcatus* Heinz; Kotsubinsky, p. 122, pl. 17, figs 5–6.
1974	*Inoceramus inaequivalvis* Schlüter; Kotsubinsky, p. 77, pl. 13, fig. 3.
1982	*Inoceramus inaequivalvis falcatus* Heinz; Keller, p. 74, pl. 4, figs 4–5.

Type. The lectotype is the specimen illustrated by Goldfuss (1934–1940, pl. 112, fig. 2d–e).

Material. Eleven specimens from the basal *Mytiloides incertus* Zone (topmost *Scaphites whitfieldi* and basal *Scaphites nigricollensis* zones) of the North Belle Fourche section, South Dakota, localities 29 (21195) and 30 (21196); nine left and two right valves.

Dimensions (*mm*).	h	l	H	L	s	VR	α	δ	B	h max
USNM 501228	62	43	57	47	34	44	78	58	29	75
USNM 510233	62	40	—	—	—	47	79	66	26	78

EXPLANATION OF PLATE 7

Figs 1–2, 6–7. *Mytiloides incertus* (Jimbo, 1894). 1, USNM 501268, USGS Mesozoic locality D12671 (Text-fig. 11, loc. 86). 2, USNM 501263; 6, USNM 501264: USGS Mesozoic locality D12671 (Text-fig. 11, loc. 86). 7, USNM 501265, USGS Mesozoic locality 20939 (Text-fig. 11, loc. 22).

Figs 3. *Inoceramus* aff. *dakotensis* sp. nov., USNM 501258, USGS Mesozoic locality D7252 (Text-fig. 11, loc. 68).

Figs 4–5. *Inoceramus dakotensis* sp. nov. 4, USNM 501271; 5, USNM 501290: USGS Mesozoic locality D7002 (Text-fig. 11, loc. 84).

All × 1.

PLATE 7

WALASZCZYK and COBBAN, *Inoceramus, Mytiloides*

USNM 501232	32	25·5	34	25·5	19	23	105	75	19	44
USNM 501234	43	27	—	—	—	31·5	—	—	19	—

Description. Moderate sized, inequilateral, strongly inequivalved. Outline elongate ovate. LV strongly inflated (B/h ratio approximating 50 per cent) with slender umbonal part, beak curved antero-dorsally, markedly projecting above hinge line. Anterior face straight, high and steep; very long (VR/h being 60–70 per cent); ventral margin narrowly rounded; posterior margin short, concave. Hinge line moderately long, straight (s/h usually about 50 per cent). Posterior auricle well-separated from disc along well-developed auricular sulcus.

RV markedly smaller and less inflated (B/h in measured specimens *c*. 25 per cent), prosocline, with subquadrate outline. Anterior face straight; ventral margin widely rounded. Hinge line straight, long (s/h above 50 per cent).

Adult ornament, in both valves, composed of widely spaced, low, round-topped, weakly developed, subregular rugae. Juveniles smooth.

Ligament massive, with high ligamental plate. Resilifers rectangular, markedly higher than long (height/length ratio more than 4) with narrow inter-resilifer areas. Measured height of ligamental plate in USNM 501233 is 7 mm. Parts of nacreous layer preserved in all specimens.

Remarks. Three subspecies of *Inoceramus inaequivalvis* have been recognized in Europe: *I. inaequivalvis inaequivalvis* Schlüter, 1877, *I. inaequivalvis modestus* Heinz, 1932 (Heinz 1932*a*), and *I. inaequivalvis falcatus* Heinz 1928 (Heinz 1928*a*). Tröger (1967, pp. 79–84) distinguished *I. inaequivalvis inaequivalvis* and *I. inaequivlavis modestus* which, based on his range chart (Tröger 1967, enclosure 43), could be regarded as two chronosubspecies, with the nominate subspecies ranging through the Middle Turonian and subspecies *modestus* occurring within the basal Upper Turonian. Subspecies *modestus* and *falcatus* were synonymized by Keller (1982) who referred both forms to *Inoceramus inaequivalvis falcatus*. According to Tröger (1967, pp. 82–83), subspecies *modestus* (=*falcatus* according to Keller 1982) possesses a more inflated RV (although being simultaneously less inflated in its LV), is more elongated axially (has higher l/h ratio), and has a smaller beak-angle than the nominate subspecies. The problem is, however, that all these characters are quite variable even in our relatively small sample (six specimens). The species varies markedly also in the material from the *Holaster planus* Zone of England, as illustrated by Woods (1912, pl. 52, figs 4–6), as well as in collections from south-eastern Europe, where *I. inaequivalvis* is relatively common. The variability is great enough to allow recognition of any of the described subspecies in any of the samples. Thus, although further material may reveal a phyletic succession within *I. inaequivalvis*, and allow recognition of separate chronosubspecies, criteria applied to separate *inaequivalvis*, *modestus*, and *falcatus* subspecies are not sufficient to justify their use in terms of rigorous taxonomic procedure, although these names may be used in a descriptive sense in distinguishing between particular morphotypes.

Occurrence. So far known only from the basal *M. incertus* Zone (topmost *S. whitfieldi* ammonite Zone) of the North Belle Fourche section in the Western Interior. This is the first record of Schlüter's species from the Western Interior, although Cobban (*in* Tourtelot and Cobban 1968) noted inoceramids closely resembling *I. inaequivalvis* in the phosphatic nodules at the base of the Niobrara Formation on the east flank of the Black Hills. The species is fairly common in the basal Upper (?and upper Middle) Turonian in Europe.

EXPLANATION OF PLATE 8

Figs 1, 4–6. *Inoceramus dimidius* White, 1874. 1, USNM 501199, USGS Mesozoic locality 21187 (Text-fig. 11, loc. 23). 4, USNM 501211, USGS Mesozoic locality 21189 (Text-fig. 11, loc. 25). 5, USNM 501206, USGS Mesozoic locality 21187 (Text-fig. 11, loc. 23). 6, USNM 501207, USGS Mesozoic locality 21189 (Text-fig. 11, loc. 25).

Fig. 2. *Inoceramus* cf. *cuvieriformis* Pergament, 1971, USNM 501442, USGS Mesozoic locality 21196 (Text-fig. 11, loc. 30).

Fig. 3. *Inoceramus* cf. *hobetsensis* Nagao and Matsumoto, 1939, USNM 501444, USGS Mesozoic locality 21196 (Text-fig. 11, loc. 30).

Fig. 7. *Inoceramus* cf. *stuemckei* Heinz, 1928, USNM 501443, USGS Mesozoic locality 21196 (Text-fig. 11, loc. 30). All ×1.

PLATE 8

WALASZCZYK and COBBAN, *Inoceramus*

Inoceramus cf. *stuemckei* Heinz, 1928

Plate 8, figure 7; Plate 9, figure 5; Plate 10, figure 7

cf. 1912 *Inoceramus lamarcki* var. *Cuvieri* Sowerby; Woods, p. 323, Text-fig. 82.
cf. 1928*a* *Inoceramus latus* Mantell, var. *stümckei* Heinz, p. 72.
cf. 1939 *Inoceramus iburiensis* Nagao and Matsumoto, p. 291, pl. 32, fig. 2.
cf. 1967 *Inoceramus lamarcki stümckei* Heinz; Tröger, p. 67, pl. 5, fig. 9; pl. 6, fig. 6 (with synonymy).
cf. 1975 *Inoceramus hobetsensis* Nagao and Matsumoto; Noda, p. 249, pl. 33, fig. 4.

Description. We have three specimens that resemble Heinz' species very closely. All come from locality 30 (21196) in the Black Hills area. All are relatively large, weakly inflated, upright forms. They possess a long and straight anterior margin, and a wide, rounded ventral margin. Hinge line long, straight. Posterior auricle (preserved only in two smaller specimens) large, extended posteriorly, well separated from disc. Ornament composed of regular, widely spaced, asymmetrical rugae, with ventral sides steeper. Concentric rugae not visible on surface of posterior ear. Rugae slightly subpentagonal in outline, as is well seen on largest specimen (Pl. 8, fig. 7).

Remarks. *I. stuemckei* Heinz is a common form in the Middle Turonian of Europe. It also occurs frequently in the lower part of the Upper Turonian, where large representatives dominate the inoceramid record and co-occur with first representatives of *I. perplexus* (= *I. costellatus* of authors) (see Ernst *et al.* 1983).

All specimens here referred to Heinz' *I. lamarcki stuemckei* very closely resemble, however, the upright, auriculate representatives of *I. hobetsensis* Nagao and Matsumoto (Noda 1975, pl. 33, fig. 4; Noda and Matsumoto 1998, pl. 16, fig. 1). The Japanese species differs only in possessing a radial sulcus, although Noda (1975) stated that this is not a taxonomically significant feature of *I. hobetsensis*.

Occurrence. *I. stuemckei* Heinz commonly occurs in the Middle–lower Upper Turonian of Europe, being known from Russia, Poland, Germany, the Czech Republic, France, and England. The present report from the lower *incertus* Zone of the Belle Fourche section, South Dakota, is the first record from the US Western Interior.

Inoceramus cf. *cuvieriformis* Pergament, 1971

Plate 8, figure 2

cf. 1822 *Inoceramus Cuvieri* Mantell, p. 213, pl. 28, fig. 4.
cf. 1912 *Inoceramus Lamarcki* Parkinson; Woods, p. 307, text-fig. 69.
cf. 1971 *Inoceramus cuvieriformis* Pergament, p. 52, pl. 6, figs 1–2; pl. 7, figs 1, 5.
cf. 1971 *Inoceramus* cf. *koegleri* Andert; Pergament, pl. 15, fig. 1.

Description. We have a single incomplete RV mould. Juvenile part with weathered surface. Valve inequilateral with axially elongated outline, moderately inflated, prosocline. Umbonal region pointed, projecting above hinge line. Posterior auricle well separated from disc, moderately large, flat. Anterior margin long, straight; anterior face high, steep. Ventral margin narrowly rounded. Posterior margin slightly concave. Hinge line moderately long, straight. Valve ornamented by regular, asymmetrical, round-topped rugae, regularly increasing in size and spacing ventralward, passing onto posterior auricle.

EXPLANATION OF PLATE 9

Figs 1–3. *Mytiloides bellefourchensis* sp. nov. 1, USNM 501249, USGS Mesozoic locality 21295 (Text-fig. 11, loc. 29). 2, USNM 501267, USGS Mesozoic locality D7002 (Text-fig. 11, loc. 84). 3, USNM 501248, USGS Mesozoic locality 21194 (Text-fig. 11, loc. 28).
Fig. 4. *Inoceramus dakotensis* sp. nov., USNM 501266, USGS Mesozoic locality D7002 (Text-fig. 11, loc. 84).
Fig. 5. *Inoceramus* cf. *stuemckei* Heinz, 1928, USNM 501255, USGS Mesozoic locality 21196 (Text-fig. 11, loc. 30).
Fig. 6. *Inoceramus sachsi* Bodylevski, 1958, USNM 501254, USGS Mesozoic locality 21765 (Text-fig. 11, loc. 34).
Fig. 7. *Inoceramus perplexus* Whitfield, 1877, USNM 501269, USGS Mesozoic locality D7002 (Text-fig. 11, loc. 84).
All × 1.

PLATE 9

WALASZCZYK and COBBAN, *Inoceramus, Mytiloides*

Remarks. Our specimen resembles particularly the specimen illustrated by Pergament (1971, pl. 15, fig. 1), referred by him to *Inoceramus* cf. *koegleri* Andert. Andert's form has a different outline and possesses quite different ornament, characteristic for the *Cremnoceramus-Tethyoceramus* representatives, whereas Pergament's specimen has a typical *lamarcki* ornament. It should, most probably, be included in the synonymy of his species *I. cuveriformis*.

Occurrence. Basal *incertus* Zone of the Belle Fourche section.

Inoceramus cf. *hobetsensis* Nagao and Matsumoto, 1939

Plate 8, figure 3

compare:

1939 *Inoceramus hobetsensis* Nagao and Matsumoto, p. 281, pl. 28, fig. 3; pl. 29, figs 1–6; pl. 30, figs 2–3.
1939 *Inoceramus hobetsensis* var. *nonsulcatus* Nagao and Matsumoto, p. 282, pl. 27, fig. 3; pl. 28, fig. 4; pl. 30, fig. 1.
1975 *Inoceramus hobetsensis* Nagao and Matsumoto; Noda, p. 249, pl. 32, figs 6–9; pl. 33, figs 1–7; pl. 34, figs 1–5; pl. 35, fig. 1 (with synonymy).
1992 *Inoceramus hobetsensis* Nagao and Matsumoto; Elder and Box, p. 19, figs 10.7–10.8, 10.10–10.13, 11.1–11.6.
1999 *Inoceramus* (*Inoceramus*) *hobetsensis* Nagao and Matsumoto; Noda and Matsumoto, p. 439, pl. 16, figs 1–3.

Type. The lectotype, designated subsequently by Noda (1975, p. 249), is the unregistered specimen, housed in the Hokkaido University Collection, illustrated by Nagao and Matsumoto (1939, pl. 29, fig. 3; re-illustrated by Noda and Matsumoto 1999, pl. 16, fig. 2) from the Middle Yezo Group of the Hobetsu area, south-central Hokkaido.

Remarks. We have a single incomplete RV from the basal *M. incertus* Zone of the Black Hills area. It lacks the juvenile part as well as the ventral part. The concentric rugae and axial sulcus on the disc very closely resemble those of *I. hobetsensis*.

Three morphotypes can be distinguished in *I. hobetsensis* Nagao and Matsumoto: (1) the typical form, with axial radial sulcus; (2) a form without a radial sulcus (var. *nonsulcatus* of Nagao and Matsumoto 1939, p. 282); and (3) a nodose form, identical to type 1 but for the presence of swelling on the rugae along the margins of the radial sulcus. Kauffman (1977*a*, p. 177) compared the latter morphotype with the American species *Inoceramus flaccidus* White, 1876, described originally from the Middle Turonian near Pueblo, south-east Colorado (White 1877, pl. 16, fig. 1, re-illustrated by Stanton 1894, pl. 13, fig. 1; Logan 1898, pl. 90; Kauffman 1977*b*, pl. 8, fig. 5). White's species is, however, much less regular in outline, and the prominent, well-developed radial sulcus is in a much more posterior position than in *I. hobetsensis*.

EXPLANATION OF PLATE 10

Figs 1–2, 4. *Mytiloides mytiloidiformis* (Tröger, 1967). 1, holotype (plaster cast of original of Tröger 1967, pl. 11, fig. 4; collections of Bundesanstalt für Geowissenschaften und Rohstoffe, Berlin, Germany). 2, USNM 501449, USGS locality D2978 (Text-fig. 11, loc. 71). 4, USNM 501448, USGS Mesozoic locality 22953 (Text-fig. 11, loc. 54).
Fig. 3. *Mytiloides aviculoides* (Meek and Hayden, 1860), plaster cast of the lectotype, USNM 242b.
Fig. 5. *Mytiloides incertus* (Jimbo, 1894), USNM 501221, USGS Mesozoic locality 21195 (Text-fig. 11, loc. 29).
Fig. 6. *Inoceramus* ex gr. *lamarcki* Parkinson, 1818, USNM 501256, USGS Mesozoic locality 21765 (Text-fig. 11, loc. 34).
Fig. 7. *Inoceramus* cf. *stuemckei* Heinz, 1928, USNM 501257, USGS Mesozoic locality 21196 (Text-fig. 11, loc. 30).
All ×1.

WALASZCZYK and COBBAN, *Inoceramus, Mytiloides*

Occurrence. *I. hobetsensis* is known from the Middle and lower Upper Turonian in the North Pacific region (Japan, Russian Far East, Alaska) and in the Western Interior.

Inoceramus sachsi Bodylevski, 1958

Plate 9, figure 6; Text-figure 14

1958 *Inoceramus sachsi* Bodylevski, *in* Bodylevski and Shulgina, p. 79, pl. 32, fig. 1; pl. 33, fig. 1.
non 1958 *Inoceramus* cf. *sachsi* Bodylevski, *in* Bodylevski and Shulgina, p. 79, pl. 30, fig. 1.
1971 *Inoceramus sachsi* Bodylevski; Pergament, p. 83 (*pars*), pl. 19, fig. 1.

Type. Holotype by original designation is the original of Bodylevski (*in* Bodylevski and Shulgina 1958, pl. 32, fig. 1) from the Turonian of Taymyr, Western Siberia.

Description. We have two specimens represented by single, relatively well-preserved LV and RV moulds, 110 and 130 mm high (h max), although the specimens could have been even twice as large when complete. Both specimens are moderately inflated (B/h approximately 0·3) with a straight, very long anterior margin, slightly concave near umbo (VR consisting up to 80 per cent of the respective axial length), and steep, high, anterior face. As well seen in RV (USNM 501463), anterior face forms distinct, flat and smooth area, with very clear boundary with lateral part of disc. Ventral margin rounded, passing into broadly convex posterior margin. Hinge line not preserved, inferred to have been relatively long. Umbonal part with pointed beak projecting above hinge line, curved antero-dorsally. Ligament very thick, estimated height up to 2 cm. Disc covered with regular, round-topped, strong rugae with wide inter-rugae spaces, increasing gradually towards the ventral margin. Rugae poorly visible up to 30 mm axial length, with shell in that part almost smooth. Some indistinct radial elements are seen in the USNM 501463.

Remarks. Our specimens are identical to the specimens from Taymyr (northern Siberia, Russia), illustrated by Bodylevski (*in* Bodylevski and Shulgina 1958, pl. 32, fig. 1; pl. 33, fig. 1). A very similar specimen was illustrated by Pergament (1971, pl. 19, fig. 1) from Kamchatka (Pacific Russia).

The specimens described here are similar in general outline to *Inoceramus iburiensis* Nagao and Matsumoto (e.g. Nagao and Matsumoto 1939, pl. 22, fig. 2). They differ, however, in lacking a radial sulcus, and in having stronger and more regular ornament.

Occurrence. The Upper Turonian specimen USNM 501254 comes from the topmost part of the *incertus* Zone of the Belle Fourche section. USNM 50146 is from the lower part of the *incertus* Zone of the same section.

I. sachsi was described from Kamchatka by Pergament (1971) who reported it from his *Inoceramus verus* Zone, which he placed in the Lower Coniacian. However, the top of his *verus* Zone is marked by the FAD of *I. multiformis* Pergament, 1971, a form that is regarded as a biogeographical variant of *I. teshioensis*, an Upper Turonian index in Japan and Sachalin (Zonova 1992; Zonova and Yazykova 1999). Consequently *I. sachsi* from Kamchatka is of Late and possibly even of latest Middle Turonian age.

Similarly the Coniacian age of *I. sachsi* from Taymyr is very problematic. It was given in an original paper by Bodylevski (*in* Bodylevski and Shulgina 1958) and recently by Khomentovsky (1998). Both authors placed the base of the Coniacian at the appearance level of *Inoceramus subinvolutus*, which they regarded as closely related to the Middle Coniacian *Volviceramus involutus* (Sowerby, 1829). Based on the illustration in Bodylevski (*in* Bodylevski and Shulgina 1958, pl. 33, fig. 3), it is an example of *I. inaequivalvis* Schlüter, 1877, a common early Late Turonian form in Europe and also known from the Western Interior (this paper). The base of the succeeding zone of *I. russiensis* Nikitin, 1888, is equated with the upper part of the Middle Coniacian of Europe (Khomentovsky 1998). The main interval with forms which may be ascribed to *I. russiensis* Nikitin occurs, however, at the top of the Lower Coniacian (see Walaszczyk and Wood, *in* Niebuhr *et al.* 1999). The chronostratigraphical position of particular zones from Taymyr as given by Bodylevski (*in* Bodylevski and Shulgina 1958) and Khomentovsky (1998) thus seems to be too high. Consequently, the stratigraphical position of forms referred to *I. sachsi* from that area may actually be distinctly lower chronostratigraphically, well below the Coniacian.

TEXT-FIG. 14. *Inoceramus sachsi* Bodylevski, 1958, USNM 501463, USGS Mesozoic locality 21196 (Text-fig. 11, loc. 30); ×0.7.

Inoceramus annulatus Goldfuss, 1836

Plate 31, figure 3

1836 *Inoceramus annulatus* Goldfuss, p. 114 (*pars*), pl. 110, fig. 7a (*non* fig. 7b).
1926 *Inoceramus* ex aff. *annulatus* Goldf.; Heinz, p. 99.
1928*c* *Inoceramus annulatus* Goldf., part. Heinz; Heinz, p. 73, pl. 5, fig. 2.
1958 *Inoceramus* cf. *annulatus* Goldfuss; Kotsubinsky, p. 14, pl. 5, fig. 23.
1962 *Inoceramus annulatus* Goldfuss; Radwańska, p. 149, pl. 5, fig. 3; pl. 6, figs 1–2.
?1962 *Inoceramus circularis* Schlüter; Radwańska, p. 150, pl. 7, fig. 1; pl. 8, fig. 1.
?1965 *Inoceramus annulatus* Goldfuss; Arzumanova, p. 121, pl. 1, fig. 3.
1968 *Inoceramus annulatus* Goldfuss; Kotsubinsky, p. 129, pl. 19, figs 3–4.
?1988 *Inoceramus* cf. *annulatus* Goldfuss; Tröger, *in* Tröger and Christensen, p. 28, pl. 4, fig. 1.
1999*a* *Inoceramus annulatus* Goldfuss; Walaszczyk and Wood, pl. 3, fig. 2; pl. 18, fig. 3.

Type. The lectotype by the subsequent designation of Heinz (1926, p. 99), is the original of Goldfuss (1836, pl. 110, fig. 7a), from the '*Scaphitensschichten*' (?Upper Turonian–?Lower Coniacian) of northern Germany.

Material. Two specimens from the middle Lower Coniacian of locality 16 (21421) of Shelby, Montana.

Remarks. This species was discussed recently by Walaszczyk and Wood (1999*a*). The two American specimens are a very typical upright form, with very regular ornament of asymmetrical, widely spaced rugae that are almost circular in outline.

Occurrence. The species appears in the Lower Coniacian *Cremnoceramus waltersdorfensis hannoverensis* Zone and ranges up to the '*Inoceramus gibbosus*' Zone of the uppermost Lower Coniacian in Europe (see Niebuhr *et al*. 1999). The Shelby specimens are from the Lower Coniacian *Cremnoceramus erectus* Zone.

Inoceramus longealatus Tröger, 1967

Text-figure 15

1930 *Inoceramus costellatus* Woods; Fiege, p. 35 (*pars*), pl. 5, figs 10–11 (non pl. 5, figs 3–9).
1967 *Inoceramus vancouverensis longealatus* Tröger, p. 95, pl. 10, fig. 2.
non 1974 *Inoceramus* cf. *vancouverensis longealatus* Tröger; Sornay, p. 31, pl. 2, fig. 2.
non 1978 *Inoceramus longealatus* Tröger, n. subsp. (with coarse ribs); Wiedmann and Kauffman, pl. 2, fig. 12.
?*non* 1982 *Inoceramus costellatus longealatus* Tröger; Keller, p. 94, pl. 7, fig. 2.
?1986 *Inoceramus parvus* Tröger; Scott *et al.*, fig. 6f.
non 1986 *Inoceramus longealatus* Tröger; Scott *et al.*, fig. 6a–e, i.
1991 *Inoceramus longealatus* Tröger; Collom, pl. 1, figs 2, 5.
non 1992 *Inoceramus longealatus* Tröger; Elder and Box, p. 21, figs 1–4, 6–8.
1997 *Inoceramus longealatus* Tröger; Walaszczyk and Szász, p. 776, figs 5c–d, 6d–h.

Holotype. By original designation, this is specimen 815 illustrated by Tröger (1967, pl. 10, fig. 2) from the Upper Turonian of Hoppenstedt, Germany, which is housed in the collections of the Geological Institute of the Mining Academy, Freiberg.

Material. Three specimens: USNM 501445 (Text-fig. 15A) from locality 79 (8978); USNM 501446 (Text-fig. 15B) and USNM 501447 (not illustrated) from locality 75 (D13433).

Dimensions (*mm*).	h	l	H	L	s	VR	α	δ	B	h max
USNM 501445	33·5	36·5	33	41	31	29	125	75	—	41
USNM 501446	37	38	37	51	38	28	128	80	—	51·5
USNM 501447	28	36	29·5	33	27	26	—	83	—	40

Description. Small to moderate-sized, inequilateral, ?equivalved. Anterior margin moderately long (VR/h = 60 per cent), straight or slightly convex; ventral margin widely rounded. Posterior margin concave. Hinge line very long, straight (up to 100 per cent of the respective axial length). Valve outline subquadrate. Posterior auricle widely extended posteriorly, well separated from disc along auricular sulcus.

Valves ornamented by regularly to subregularly spaced, raised growth lines and superimposed sub- to irregular, low, often indistinct rugae. Ornament also well developed on surface of posterior auricle, forming sickle-shaped lobe in dorsal part, meeting hinge line at right or even smaller angle.

Remarks. *Inoceramus longealatus* is a very difficult species to recognize. An essentially identical posterior auricle may occur in other, totally unrelated species, e.g. in some of the late Turonian mytiloids, as well as in *Inoceramus dakotensis* (see above), juveniles of which are, in practice, indistinguishable from Tröger's type. In Western Interior material from the upper Upper Turonian (upper *incertus* and basal *scupini* Zones), there are small, subquadrate specimens that appear to be adult (Text-fig. 15); these match exactly the characteristics of *I. longealatus*. Identical forms were also reported from the upper Upper Turonian of Romania (Walaszczyk and Szász 1997, figs 5c–d, 6d, h).

Relative stratigraphical position and morphological resemblance suggest an evolutionary relationship between *I. longealatus* and *I. dakotensis*. Some phyletically advanced specimens of the latter display a widely extended posterior auricle, and an ornament closely resembling that of Tröger's species.

The specimens referred to *I. longealatus* by Scott *et al.* (1986) are distinctly elongated mytiloids, here referred to *Mytiloides scupini* (Heinz, 1930) (see also Walaszczyk and Tröger 1996). The specimen illustrated and placed by Scott *et al.* (1986, fig. 6f) in *Inoceramus parvus* Tröger, 1981, should probably be referred to *I. longealatus*.

Specimens illustrated as *I. longealatus* by Elder and Box (1992) from the Turonian of south-western Alaska look very heterogeneous, and none represents Tröger's species. Their specimen USNM 452180 (Elder and Box 1992, fig. 12.1) is quite distinct from the others, and is conspecific with two of their specimens referred to *Inoceramus frechi* (Elder and Box's figs 12.9–12.10 only), although what they call *I. frechi* seems to be closer to the North Pacific *uwajimensis* group (see Nagao and Matsumoto 1939, pl. 33;

TEXT-FIG. 15. *Inoceramus longealatus* Tröger, 1967. A, USNM 501445, USGS Mesozoic locality 8978 (Text-fig. 11, loc. 79); B, USNM 501446, USGS Mesozoic locality D13433 (Text-fig. 11, loc. 75): both ×1.

Zonova 1970; Noda 1975), and differs from Flegel's *I. frechi* (see Scupin 1912–1913; Walaszczyk and Tröger 1996). Three other specimens referred to *I. longealatus* in their figure 12.2–4 are small, subrounded forms with a sulcus in the posterior part of the disc. They differ from *I. longealatus*, moreover, in valve outline and character of the posterior auricle, which in Alaskan specimens is small and not well-extended posteriorly. The remaining two specimens referred to *I. longealatus* (their fig. 12.6–8) are, in contrast to upright *I. longealatus*, markedly oblique. Finally, all specimens illustrated by these authors differ from *I. longealatus* in ornament. Tröger's species possesses *Mytiloides*-like ornament, with fine, raised growth lines and low, widely spaced rugae.

The specimen referred to as '*I. longealatus* Tröger, n. subsp. with coarse ribbing' by Wiedmann and Kauffman (1978, pl. 2, fig. 12), represents a quite different species. Judging from their illustration, what was to be an extended posterior auricle, characteristic for *I. longealatus*, is not an auricle, but the posterior shell of a non-auriculate individual, with subcircular rugae curved toward the beak in the dorsal part.

Occurrence. Rare in the *incertus* and *scupini* zones in Colorado and New Mexico, in the US Western Interior. Known from the same stratigraphical interval in Germany, Poland, Russia, Romania, and the Czech Republic in Europe.

Genus MYTILOIDES Brongniart, 1822

Type species. By monotypy, *Ostracites labiatus* Schlotheim, 1813, p. 93.

Remarks. See Kauffman and Powell (*in* Kauffman *et al.* 1977), Harries *et al.* (1996), and Walaszczyk and Wood (1999*a*) for discussion.

Representatives of the genus *Mytiloides* are the dominant element of Upper Turonian inoceramid assemblages in both the Western Interior and Europe. Although all the *Mytiloides* species recognized here are known also from the Old World, the Western Interior record is characterized by the dominance of *M. incertus* (Jimbo, 1894) and allied forms: *M. mytiloidiformis* (Tröger, 1967), *M. ratonensis* sp. nov., and *M.* aff. *ratonensis* sp. nov. Other species that may be referred informally to the *striatoconcentricus* group, which dominates the European record, are much less common.

Occurrence. Uppermost Cenomanian–Lower/?Middle Coniacian world-wide.

Mytiloides incertus (Jimbo, 1894)

Plate 7, figures 1–2, 6–7; Plate 10, figure 5; Plate 11, figure 3; Plate 12, figures 1–12; Plate 13, figures 1–8, 10; Plate 14, figure 5; Text-figure 16

1894 *Inoceramus incertus* Jimbo, p. 189, pl. 24, fig. 7.
1940 *Inoceramus incertus* Jimbo em.; Nagao and Matsumoto, p. 10 (*pars*), pl. 3, figs 1–3; pl. 10, fig. 2 (*non* pl. 3, fig. 4).
?1940 *Inoceramus* cf. *incertus* Jimbo em.; Nagao and Matsumoto, p. 10, pl. 3, fig. 5.
1963 *Inoceramus incertus* Jimbo; Matsumoto *et al.*, pl. 67, fig. 7.
1967 *Inoceramus fiegei fiegei* Tröger, p. 105, pl. 11, fig. 3; pl. 13, figs 14–15, 17, 20 (with synonymy).
1974 *Inoceramus striatus* Mantell; Kotsubinsky, p. 78, pl. 14, fig. 2.
?1976 *Inoceramus fiegei fiegei* Tröger; Lupu, p. 133, pl. 1, fig. 4.
1977*b* *Mytiloides fiegei fiegei* (Tröger); Kauffman, pl. 11, fig. 1; pl. 12, fig. 4.
1978 *Mytiloides fiegei fiegei* (Tröger); Kauffman *et al.*, pl. 15, fig. 1; pl. 16, fig. 4.
1982 *Inoceramus fiegei fiegei* Tröger; Keller, p. 110, pl. 7, fig. 3.
1983 *Mytiloides incertus* (Jimbo); Matsumoto and Noda, p. 109, figs 2–4.
1984 *Mytiloides incertus* (Jimbo); Noda, p. 458, text-figs 7–8, pls 84–85; pl. 86, figs 1–8 (with synonymy).
1991 *Mytiloides* aff. *mytiloidiformis* (Tröger); Collom, pl. 2, fig. 1.
1991 *Mytiloides incertus* (Jimbo); Collom, pl. 2, figs 2, 4.
1998 *Mytiloides incertus* (Jimbo); Lucas and Estep (*pars*), fig. 5F–G.

Type. The lectotype by the subsequent designation of Matsumoto and Noda (1983) is MM7535, one of Jimbo's unfigured syntypes, in a pebble from the River Pombets, Mikasa City, central Hokkaido. The specimen is housed in the University Museum in Tokyo.

Material. Numerous specimens in the collection of the Geological Survey in Denver.

Dimensions (*mm*).	h	l	H	L	s	l at H = 35	α	δ	ribs in h distance 10–25	15–35
USNM 501291	33	25	27	29·5	12	30·5	91	48	4	5
USNM 501280	62	44·5	50	50	23·5	27	98	50	4	5
USNM 501263	42	32·4	35	35	18	27	98	50	5	7
USNM 501275	38	31	31·2	33·5	16	28	96	50	5	5
USNM 501264	31	25·6	27·2	27	12·2	—	103	58	4	4
USNM 501306	44	34·4	40	41	22·2	29.5	98	56	3	4
USNM 501262	44	33	37·7	36·3	16·3	26·5	98	58	3	4
USNM 501287	38·2	27·2	30·5	33	15	26·5	104	50	4	4
USNM 501290	38·4	27	31·4	32·5	15·2	27	90	50	4	5
USNM 501288	34	26·5	30	29	16·5	26	—	—	5	6
USNM 501289	53	33	46·7	37·4	20·5	23	100	60	4	4
USNM 501292	45	38	37	40	24	30·5	100	50	3	4
USNM 501294	47·5	36	41·5	39	18	27·5	110	54	5	5
USNM 501295	49·5	37·4	46·4	42	17	26·5	112	50	5	5
USNM 501297	48	37·2	43	39	17	27	105	60	4	5
USNM 501286	32	28·6	29	29·6	11	31	114	52	4	5
USNM 501296	50	40	41	42	21	30	107	49	4	5
USNM 501298	40	31·5	36·5	34·2	17	27·2	104	53	5	7
USNM 501285	32	27·5	29	30	17	28	107	54	4	4

Description. Small to medium-sized for genus, inequilateral, equivalve, weakly inflated, with maximum inflation dorso-central. Valve outline subrounded to ovate, elongated parallel to growth axis, markedly oblique (with angle δ usually between 50 and 60 degrees). Anterior margin convex, rarely straight, short, usually less than half of corresponding axial length. Anteroventral margin long, broadly convex, Ventral margin rounded, passing into straight or slightly convex posterior margin. Hinge line straight, relatively short. Anterior face steep, rounded. Posterior auricle small, well separated from disc. Growth axis straight or slightly convex anteriorly.

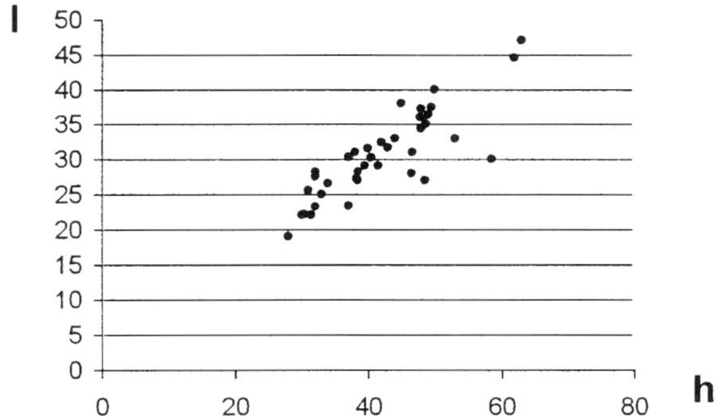

TEXT-FIG. 16. l/h ratio in selected specimens of *Mytiloides incertus* (Jimbo, 1894).

Valves covered with regularly, subevenly spaced concentric rugae, with inter-rugae spaces covered with distinct, raised growth lines. Ornament changes in adult part to more closely spaced round-topped rugae.

Remarks. *Mytiloides incertus* displays a relatively high range of variability, as was thoroughly discussed by Noda (1984). This variability chiefly affects the valve outline and ornament (ribbing density), spanning subrounded forms with widely spaced, distinct rugae (Pl. 14, fig. 5), to axially elongated forms with either widely (Pl. 11, fig. 3; Pl. 12, figs 10, 12; Pl. 13, figs 4–5, 8), or closely spaced rugae (Pl. 12, figs 6, 11; Pl. 13, figs 2, 6–7). Some juveniles with relatively closely spaced rugae (e.g. Pl. 13, figs 7, 10) may be inseparable from juveniles of *M. ratonensis* sp. nov. (see Pl. 13, fig. 9; Pl. 15, fig. 10).

In contrast to Japanese and European material, North American representatives of *M. incertus* possess sharp-edged rugae, with wide, flat-floored interspaces (see Pl. 11, fig. 3; Pl. 12, figs 3–5, 10, 12; Pl. 13, figs 4–5, 8). They resemble the specimen illustrated by Nagao and Matsumoto (1940, pl. 3, fig. 3), which Noda (1984, p. 463) regarded as still within the variability range of the species. However, in North American material such forms dominate the record and individuals comparable in respect of surface ornament to those illustrated from Japan are extremely rare. The North American assemblage may well be a geographical subspecies; study of material from the Gulf Coast is required to clarify the geographical distribution of these two morphotypes.

M. fiegei fiegei (Tröger, 1967) is clearly a junior synonym of *Mytiloides incertus*. Kauffman's (1977*a*, p. 180) proposal of *M. incertus* as an evolutionary link between *Mytiloides fiegei fiegei* (Tröger, 1967) and '*Inoceramus*' *waltersdorfensis* Andert should be rejected as *fiegei fiegei* and *incertus* come from equivalent time intervals.

A small specimen possessing closely spaced, round-topped rugae, illustrated by Nagao and Matsumoto (1940, pl. 3, fig. 5), and referred by Kauffman (1977*a*) to *M. labiatoidiformis* (Tröger, 1967), differs from Tröger's species in general outline and in the type of rugae. On the other hand this small specimen may represent a juvenile *M. ratonensis* sp. nov., the juveniles of which are closely rugate, *incertus*-like forms (see Pl. 15). Further collections are needed in Japan, however, to show whether *M. ratonensis* sp. nov. succeeds the *incertus*-dominated interval.

Kauffman (1977a) referred one of Nagao and Matsumoto's specimens (1940, pl. 10, fig. 2), characterized by having a more rounded outline, to '*Inoceramus*' *meekianus* Anderson (1958, p. 101, pl. 22, figs 5–6), which he regarded, moreover, as an evolutionary link between *Mytiloides latus* (*sensu* Woods 1912, text-fig. 41) and *M. fiegei fiegei* (Tröger, 1967). Although the type of Anderson's species resembles the Japanese form in valve outline, its ornament differs. *I. meekianus* possesses sharp-edged rugae with wide inter-rugae spaces, as in North American *M. incertus*, but they cross the growth lines obliquely (*Querungsrippen* of Heinz 1928*b*). This feature has not been observed in *M. incertus* nor in allied species. *Querungsrippen* are found in some Turonian forms (e.g. *I. securiformis*; see Woods, 1912, text-fig. 78), but they are more characteristic of stratigraphically much younger inoceramids. Indeed, the type of *M. meekianus* may well be a Santonian–Campanian species.

Occurrence. *M. incertus* Zone (topmost part of the *Scaphites whitfieldi* Zone, *Scaphites nigricollensis* Zone and lower part of the *Prionocyclus quadratus* Zone in ammonite terms). Known over all of the Northern Hemisphere: Gulf Coast and Western Interior of North America, Brazil in South America, Asia (Japan, Afghanistan, Kazakhstan), and Europe (Russia, Poland, Germany, Czech Republic, Romania, France, Spain, England).

Mytiloides bellefourchensis sp. nov.

Plate 6, figures 4, 9–10; Plate 9, figures 1–3

1997 *Mytiloides striatoconcentricus* (Gümbel); Leckie *et al*. (*pars*), fig. 36J (*non* fig. 36H–I, ?K–L).

Derivation of name. After the town Belle Fourche, on north-eastern flank of the Black Hills.

Type. The holotype is USNM 501248 (Pl. 9, fig. 3).

Type locality. Belle Fourche section, north-eastern flank of the Black Hills, South Dakota, locality 28 (21194) (Text-figs 5–6, 11).

Type horizon. *I. dakotensis* Zone (top of the *S. whitfieldi* Zone in ammonite terms) of the Turner Member of the Carlile Shale; lower Upper Turonian.

Material. Numerous specimens from localities 27 (21192), 28 (21194), and 29 (21195).

Dimensions (*mm*).	h	l	H	L	s	VR	α	δ	B	h max
USNM 501267	34	25·5	29	27·5	19	13	110	50	—	40
USNM 501248	43	27	34·5	34	18	13	100	50	—	49
USNM 501240	43	23	34·5	32	15	10	102	49	—	41·5
USNM 501253	34·5	23	28	27	12	9	103	52	—	43

Diagnosis. Small to moderate in size, inequilateral, equivalve species. Valves prosocline, markedly oblique, weakly inflated unless geniculated. Juveniles ornamented with raised growth lines circular in outline; adults with growth lines and superimposed concentric rugae.

Description. Valves of small to moderate size, inequilateral, equivalve, prosocline, markedly oblique. Valves weakly inflated unless geniculated, with maximum inflation dorso-central. Growth axis straight, or slightly curved anteriorly. Beak pointed, projecting slightly above hinge line. Anterior margin short, convex, passing into long, weakly convex antero-ventral margin. Ventral margin narrowly rounded, usually passing into straight posterior margin. Hinge line straight, relatively short. Posterior auricle small or of moderate size, poorly separated from disc in non-geniculated forms.

Juvenile ornament composed of raised growth lines; adult ornament of growth lines, and superimposed regularly and closely spaced, concentric rugae.

Remarks. *M. bellefourchensis* sp. nov. differs from closely similar *Mytiloides striatoconcentricus* (Gümbel, 1868) in being weakly inflated and possessing a circular outline in the juvenile part. Moreover,

---

EXPLANATION OF PLATE 11

Figs 1–2, 4. *Inoceramus* aff. *dakotensis* sp. nov. 1, USNM 501259, USGS Mesozoic locality D7252 (Text-fig. 11, loc. 68). 2, USNM 501261, USGS Mesozoic locality 18703 (Text-fig. 11, loc. 72). 4, USNM 501260, USGS Mesozoic locality D7252 (Text-fig. 11, loc. 68).

Fig. 3. *Mytiloides incertus* (Jimbo, 1894), USNM 501262, USGS Mesozoic locality D12671 (Text-fig. 11, loc. 86). All × 1.

WALASZCZYK and COBBAN, *Inoceramus, Mytiloides*

Gümbel's species is characterized by finer ornament. Specimens from the *Scaphites whitfieldi* Zone of Mesa Verde, National Park, Colorado, referred to *striatoconcentricus* by Leckie *et al.* (1997, fig. 36J), belong to the present species.

Many of the specimens studied are distinctly geniculate, ranging from individuals with only a weak change in slope (see Pl. 6, fig. 9) to those juvenile and adult parts at almost 90 degrees (Pl. 9, fig. 1).

Occurrence. Known from the *dakotensis* Zone (upper *S. whitfieldi* ammonite Zone) of the Western Interior, as well as from the middle Upper Turonian of Europe.

Mytiloides mytiloidiformis (Tröger, 1967)

Plate 10, figures 1–2, 4; Plate 14, figures 2–3, 7–8, 10–11, 13

1930 *Inoceramus inconstans inconstans* Woods; Fiege, p. 38, pl. 6, fig. 19.
1940 *Inoceramus incertus* Jimbo; Nagao and Matsumoto, p. 10, pl. 3, fig. 4.
1967 *Inoceramus fiegei mytiloidiformis* Tröger, p. 108, pl. 11, fig. 4; pl. 13, figs 16, 18.
?1969 *Inoceramus fiegei mytiloidiformis* Tröger; Jerzykiewicz, p. 175, photo. 3.
non 1977*b* *Mytiloides fiegei mytiloidiformis* (Tröger); Kauffman, pl. 10, fig. 4.
non ?1978 *Mytiloides fiegei mytiloidiformis* (Tröger); Kauffman *et al.*, pl. 14, fig. 4.
non 1979 *Inoceramus fiegei mytiloidiformis* Tröger; Ivannikov, p. 52, pl. 9, fig. 1.
1983 *Mytiloides* sp. aff. *M. mytiloidiformis* (Tröger); Matsumoto and Noda, p. 109, fig. 5.
1984*b* *Mytiloides fiegei mytiloidiformis* (Tröger); Cobban, p. 9, pl. 1, figs 8–9, 11–12.
1984 *Mytiloides* sp. aff. *M. mytiloidiformis* (Tröger); Noda, p. 467, pl. 86, fig. 9.
1998 *Mytiloides mytiloidiformis* (Tröger); Noda and Matsumoto, pl. 15, figs 6–7.

Type. The holotype, by original designation, is the original of Fiege (1930, pl. 6, fig. 19), re-illustrated here as Plate 10, figure 1 (see also Tröger 1967, pl. 11, fig. 4) from the Upper Turonian of Lengerich, Germany. It is housed in the Museum of the Bundesanstalt für Geologie und Rohstoffe in Berlin.

Material. 15 specimens from localities 71 (D2978), 85 (D2494), 52 (D7113), 54 (22953), and 30 (21196).

Dimensions (*mm*).	h	l	H	L	s	VR	α	δ	B	h max
USNM 501449	46	24·5	41	31·5	17·4	14	115	56	—	54
USNM 501313	62·5	34	51	42	19	24	90	58	—	73
USNM 501316	26	16	23	19·5	9·5	18	100	58	—	61
USNM 501317	49·5	27	43·3	35·7	13	18	100	58	—	61

Description. A small to medium-sized, inequilateral, equivalved species. Valves markedly elongated axially (h/l approximates 50 per cent), strongly prosocline, oblique. Growth axis usually curved anteriorly. Beak pointed, projected above hinge line. Posterior auricle well separated from disc with well-developed auricular sulcus. Anterior margin short to moderately long, weakly convex, passing into widely convex anteroventral margin. Anterior face steep, rounded. Ventral margin narrowly rounded. Posterior margin long, straight. Hinge line relatively short (usually much less than 50 per cent of the corresponding axial length).

Valves ornamented with regularly, subevenly spaced, concentric, *incertus*-type rugae and superimposed growth

---

EXPLANATION OF PLATE 12

Figs 1–12. *Mytiloides incertus* (Jimbo, 1894). 1, USNM 501272. 2, USNM 501273. 3, USNM 501274. 4, USNM 501275. 5, USNM 501275. 6, USNM 501281. 7, USNM 501280. 8, USNM 501277. 9, USNM 501278. 10. USNM 501282. 11. USNM 501283. 12. USNM 501284. 1–5, 7–8, 12, USGS Mesozoic locality D12671 (Text-fig, 11, loc. 86). 9, USGS Mesozoic locality 6773 (Text-fig. 11, loc. 47). 6, USGS Mesozoic locality D12522 (Text-fig. 11, loc. 78). 10–11, USGS Mesozoic locality D8644 (Text-fig. 11, loc. 35). All ×1.

WALASZCZYK and COBBAN, *Mytiloides*

lines. Adult ornament occasionally less regular or changes into more closely spaced, round-topped, ring-like rugae (e.g. Pl. 14, fig. 11).

Remarks. Some of our specimens differ from the type of the species (re-illustrated here in Pl. 10, fig. 1), in their somewhat outwardly curved ornament on the posterior auricle (Pl. 14, figs 2–3, 7–8, 10–11, 13).

Occurrence. The species appears at the base of the *incertus* Zone (topmost *whitfieldi* Zone) and ranges apparently throughout this zone. It seems to be dominant, however, in its lower part. Known from the Western Interior and Gulf Coast of North America, Europe (England, Germany, Romania, Poland, Russia, Ukraine), and Asia (Kazakhstan, Afghanistan, Japan).

Mytiloides ratonensis sp. nov.

Plate 13, figure 9; Plate 15, figures 1–18; Plate 16, figure 9

1968 *Inoceramus labiatus opalensis* Böse forma *elongata*; Pauliuc, pl. 4, fig. 1.
1977*b* *Mytiloides dresdensis labiatoidiformis* (Tröger); Kauffman, pl. 10, fig. 2.
1977*b* *Mytiloides* (?) sp. aff. *M. kleini* (Müller); Kauffman, pl. 10, fig. 12.
1978 *Mytiloides dresdensis labiatoidiformis* (Tröger); Kauffman *et al*., pl. 14, fig. 2.
1978 *Mytiloides* (?) sp. aff. *M. kleini* (Müller); Kauffman *et al*., pl. 14, fig. 12.
1982 *Inoceramus labiatoidiformis* Keller, variant b; Keller, p. 100, pl. 5, fig. 5.
1986 *Inoceramus* aff. *labiatoidiformis* Tröger (of Keller); Scott *et al*. (*pars*), fig. 6h, (?6j).
1991 *Mytiloides* cf. *latus* (Mantell); Collom, pl. 2, fig. 3.
1997 *Mytiloides* aff. *labiatoidiformis* (Tröger); Leckie *et al*. (*pars*), fig. 37M (*non* fig. 37L).

Derivation of name. From the type locality in the Raton Basin.

Type. The holotype USNM 388264 is the original of *Inoceramus* aff. *labiatoidiformis* Tröger (*sensu* Keller) of Scott *et al*. (1986, fig. 6h) from the Fort Hays Limestone, south of Springer in the Raton Basin, New Mexico (USGS Mesozoic locality D12522).

Type locality. Outcrops south of Springer, Colfax County, New Mexico.

Type horizon. Fort Hays Limestone Member of the Niobrara Formation.

Material. Numerous specimens from USGS localities D12522, D11843, D11834, and D11981.

Dimensions (mm).	h	l	H	L	s	VR	α	δ	B	h max
USNM 501339	40	29	36·5	31	19	18	104	58	—	55
USNM 501328	43	31	36	34	23	29	112	67	—	54
USNM 501332	39	28	37	31·3	14·3	14	111	62	—	49·5
USNM 501329	38·7	29	36	30·5	16·8	20	115	59	—	45

EXPLANATION OF PLATE 13

Figs 1–8, 10. *Mytiloides incertus* (Jimbo, 1894). 1, USNM 501279, USGS Mesozoic locality D8644 (Text-fig. 11, loc. 35). 2, USNM 501299, USGS Mesozoic locality 20939 (Text-fig. 11, loc. 22). 3, USNM 501300, USGS Mesozoic locality D11843 (Text-fig. 11, loc. 69). 4, USNM 501301, USGS Mesozoic locality D9357 (Text-fig. 11, loc. 49). 5, USNM 501304, USGS Mesozoic locality D9357 (Text-fig. 11, loc. 49). 6, USNM 501302, USGS Mesozoic locality 20939 (Text-fig. 11, loc. 22). 7, USNM 501307, USGS Mesozoic locality 22953 (Text-fig. 11, loc. 54). 8, USNM 501306, USGS Mesozoic locality D12671 (Text-fig. 11, loc. 86). 10, USNM 501303, USGS Mesozoic locality D11843 (Text-fig. 11, loc. 69).

Fig. 9. *Mytiloides ratonensis* sp. nov., USNM 501308, USGS Mesozoic locality D12522 (Text-fig. 11, loc. 78).
All ×1.

WALASZCZYK and COBBAN, *Mytiloides*

Diagnosis. Small to moderate-sized, inequilateral, equivalve, flat, moderately prosocline, with axially elongated outline. Anterior margin short, convex, passing into long anteroventral, and narrowly rounded ventral margins. Posterior auricle moderate-sized, well separated from the disc. Beak pointed, projecting above the hinge line. Juvenile ornament of rounded, *incertus*-like rugae, succeeded by middle growth stage with closely-spaced, round-topped growth lines; adult stage with irregular rugae.

Description. Small to moderate-sized, prosocline, with straight to slightly convex growth axis and ovate outline. Anterior margin slightly convex, short (less than 50 per cent of the corresponding axial length), passing into long, convex, anteroventral margin. Ventral margin narrow, rounded. Posteroventral margin convex, then widely concave, passing into posteriorly elongated posterodorsal margin. Hinge line straight, moderately long to long (up to 55 per cent of the corresponding axial length). Umbo with beak distinctly projecting above the hinge line, curved anteriorly. Posterior auricle large, extended posteriorly, well separated from disc along well-developed auricular sulcus. Disc weakly inflated, with maximum inflation dorsal or dorsocentral.

Juvenile ornament consists of more or less closely spaced rugae with superimposed growth lines. Middle growth stage, usually brief, ornamented with closely spaced, fine, round-topped growth lines. Adult stage with widely and irregularly spaced sharp-edged rugae, with flat-floored interspaces. Irregularly developed growth lines often superimposed. Juvenile ornament absent from posterior auricle. Post-juvenile ornament prominent on posterior auricle with rugae and growth lines markedly curved posteriorly.

Many (but not all) specimens of *M. ratonensis* sp. nov. are clearly doubly geniculate. A positive geniculation terminates the juvenile growth stage; a negative geniculation separates middle and adult stages.

Remarks. The sequence of three ornament types, extended posterior auricle, and general valve outline, characterise this distinctive species. Variation mostly affects juvenile axial length and l/h ratio, while ornament varies from widely spaced, *incertus*-like rugae to fine, closely spaced rugae. Juveniles with closely spaced rugae resemble *Mytiloides incertus prescheri* (Tröger, 1985, p. 41, fig. 1, pl. 1), described from the Upper Turonian of Saxony, Germany, from which they may be differentiated by their characteristic outwardly extended posterior auricle.

Inoceramus labiatoidiformis Tröger, var. b. of Keller (1982, p. 100, pl. 5, fig. 5) belongs to *M. ratonensis*. It has the same ornament pattern and, although not completely preserved, clearly shows the same posterior auricle development. *Inoceramus opalensis* Seitz of Pauliuc (1968, pl. 4, fig. 1), from the Upper Turonian of Romania, also belongs here. Juveniles with relatively coarse rugae in which the posterior part of the shell is poorly preserved are indistinguishable from *M. incertus* (Jimbo, 1894).

Small, incomplete specimens of *M. ratonensis* (juveniles with partly preserved middle ornament) were illustrated by Kauffman (1977*a*, pl. 10, figs 2, 12) as *Mytiloides labiatoidiformis* (Tröger, 1967) and *Mytiloides* (?) sp. aff. *M.*? *kleini* (Müller, 1888). Both of these species have a uniform ornament throughout ontogeny. *Mytiloides* n. sp. aff. *M. fiegei* (Tröger) of Kauffman (1977*b*, pl. 10, fig. 14) may be a finely ribbed form of *M. ratonensis*. Larger specimens, with relatively small juvenile growth stage (see Pl. 16, fig. 9) closely resemble the holotype of *Mytiloides aviculoides* (Meek and Hayden, 1860) (Meek 1876, p. 63, pl. 9, fig. 4; re-illustrated here in Pl. 10, fig. 3), which lacks, however, the three ornament

---

EXPLANATION OF PLATE 14

Figs 1, 4, 6, 9, 12. *Mytiloides* aff. *M. ratonensis* sp. nov. 1, USNM 501314, USGS Mesozoic locality D7113 (Text-fig. 11, loc. 52). 4, USNM 501310, USGS Mesozoic locality D12671 (Text-fig. 11, loc. 86). 6, USNM 501315, USGS Mesozoic locality D2494 (Text-fig. 11, loc. 85). 9, USNM 501312, USGS Mesozoic locality D2978 (Text-fig. 11, loc. 71). 12, USNM 501311, USGS Mesozoic locality D11843 (Text-fig. 11, loc. 69).

Figs 2–3, 7–8, 10–11, 13. *Mytiloides mytiloidiformis* (Tröger, 1967). 2, USNM 501316, USGS Mesozoic locality D7113 (Text-fig. 11, loc. 52). 3, USNM 501309, USGS Mesozoic locality D2978 (Text-fig. 11, loc. 71). 7, USNM 501319, USGS Mesozoic locality 22953 (Text-fig. 11, loc. 54). 8, USNM 501313, USGS Mesozoic locality D2494 (Text-fig. 11, loc. 85). 10, USNM 501317; 11, USNM 501318: USGS Mesozoic locality D7113 (Text-fig. 11, loc. 52). 13, USNM 501606, USGS Mesozoic locality D7113 (Text-fig. 11, loc. 52).

Fig. 5. *Mytiloides incertus* (Jimbo, 1894), USNM 501305, USGS Mesozoic locality 6773 (Text-fig. 11, loc. 47).

All × 1.

WALASZCZYK and COBBAN, *Mytiloides*

stages of the present species and has a less outwardly curved posterior auricle. *M. aviculoides* is here interpreted as one of the youngest members of the Lower–lower Middle Turonian mytiloids (see Cobban & Reeside 1952b), and is not uppermost Turonian as claimed by Kauffman (1977*a*). Its co-occurrence with Lower–lower Middle Turonian *Mytiloides mytiloides* (=*Inoceramus problematicus* Schlotheim) was mentioned by Meek (1876).

A number of specimens from localities 71 (D2978), 85 (D2494), 69 (D11843), 86 (D12671) and 52 (D7113) (Pl. 14, figs 1, 4, 6, 9, 12), although having most of the characters of the present species, lack the middle ornament stage of closely spaced, round-topped, ring-like growth lines. They are, moreover, more slender shells; their h/l ratio ranges between 0·55 and 0·6 compared to about 0·7 in *M. ratonensis* sp. nov. These specimens clearly represent a transitional form between *M. mytiloidiformis*/?*incertus* and *M. ratonensis*. Their systematic status remains unclear, however. They may represent a separate evolutionary member of the *M. ratonensis* lineage, and thus a separate species, or may be extreme variants included in this species. Unfortunately, the specimens are not precisely located in the stratigraphical record to allow a definite statement. They are referred provisionally here to *M.* aff. *ratonensis* sp. nov.

The measurements (in mm) of some of these specimens (Pl. 14, figs 1, 4, 6, 9, 12) are as follows:

	h	l	H	L	s	VR	α	δ	B	h max
USNM 501314a	43	23	—	—	—	13	—	—	—	45
USNM 501322	39·5	26·7	34	30	17·1	11·5	106	53	—	50
USNM 501310	45·2	25	42	30·2	18·1	12·2	105	57	—	57
USNM 501324	43·8	27	—	—	—	12	—	—	—	47·4
USNM 501321	37	24·5	30·7	29	—	9	110	52	—	40·5
USNM 501323	39	26·5	37	28	13	12	110	63	—	49
USNM 501320	45	28·5	37	37	18	13	105	52	—	52
USNM 501311	36	20·5	31	24·2	16·5	10·5	100	56	—	47
USNM 501312	43	25	36·5	29	15·5	12	—	56	—	48
USNM 501315	46	26·2	38·3	31	16·5	13	100	53	—	53

Occurrence. Known from the *Mytiloides scupini* Zone of the upper Upper Turonian of the Western Interior. In Europe, known from Germany, Poland, and Romania.

Mytiloides scupini (Heinz, 1930)

Plate 16, figures 1–8, 10–11; Plate 17, figures 1–8; Plate 18, figures 10–13

1911 *Inoceramus frechi* Flegel; Andert, p. 51, pl. 1, fig. 8; pl. 7, fig. 6.
1928*a* *Inoceramus stillei* Heinz; Heinz, p. 73 (*pars*), only Andert's 1911, pl. 1, fig. 8, *I. frechi*.
1930 *Inoceramus stillei* var. *scupini* Heinz, p. 26.
1964 *Inoceramus* aff. *I. perplexus* Whitfield; Scott and Cobban (*pars*), pl. 2, fig. 1 (*non* pl. 2, figs 2–5).
1982 *Inoceramus frechi* Flegel; Keller, p. 96, pl. 7, fig. 1.
1986 *Mytiloides* aff. *mytiloidiformis* (Tröger); Cobban, fig. 6D.
1986 *Inoceramus longealatus* Tröger; Scott *et al*., figs 6a–i.
1991 *Mytiloides aviculoides* (Meek and Hayden); Collom, pl. 4, fig. 1.
1991 *Mytiloides lusatiae* (Andert); Collom, pl. 4, figs 4, 7; pl. 5, figs 1–3.

EXPLANATION OF PLATE 15

Figs 1–18. *Mytiloides ratonensis* sp. nov. 1, USNM 501325. 2, USNM 501326. 3, USNM 501327. 4, USNM 501328. 5, USNM 501329. 6, USNM 501330. 7, USNM 501331. 8, USNM 501333. 9, USNM 501334. 10, USNM 501335. 11, USNM 501332. 12, USNM 501336. 13, USNM 501337. 14, USNM 501338. 15, USNM 501339. 16, USNM 501340. 17, USNM 501341. 18, USNM 501342. 1, 9, 11, 15, 17–18, USGS Mesozoic locality D12522 (Text-fig. 11, loc. 78). 2–8, 10, 12–14, 16, USGS Mesozoic locality D11843 (Text-fig. 11, loc. 69). All ×1.

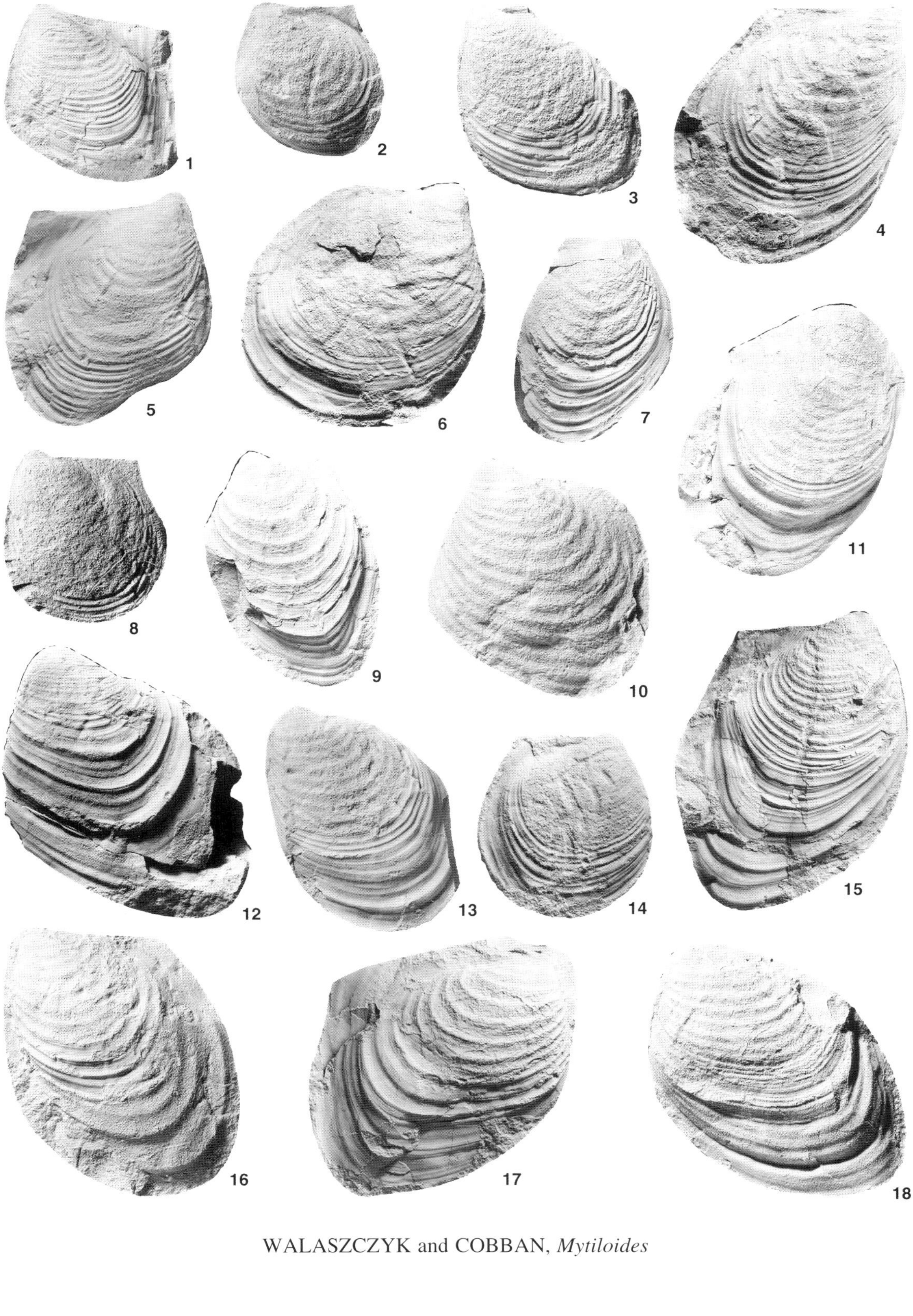

WALASZCZYK and COBBAN, *Mytiloides*

1991 *Mytiloides labiatoidiformis*? (Tröger); Collom, pl. 4, fig. 5.
1991 *Mytilodes* aff. *kleini* (Müller); Collom, pl. 4, fig. 6.
1991 *Mytilodes* aff. *aviculoides* (Meek and Hayden); Collom, pl. 5, fig. 4.
1991 *Mytiloides*? aff. *kleini* (Müller); Collom, pl. 6, fig. 6.
1992 *Mytiloides carpathicus* (Simionescu); Walaszczyk, p. 26 (*pars*), pl. 15, fig. 4 only.
1996 *Mytiloides scupini* (Heinz); Walaszczyk and Tröger, p. 400, fig. 3C–E.
1999*a* *Mytiloides scupini* (Heinz); Walaszczyk and Wood, p. 425, pl. 1, fig. 10.

Type. The holotype, by original designation (Heinz 1930, p. 30), is the original of *Inoceramus frechi* Flegel of Andert of (1911, pl. 1, fig. 8; pl. 7, fig. 6), from Sonnenberg near Waltersdorf, Germany. It is housed in the Andert collection in the State Mineralogical and Geological Museum, Dresden, Germany.

Material. Numerous specimens from USGS localities D12522, D10459, D10458, D3681, and D13433.

Dimensions (mm).	h	l	H	L	s	VR	α	δ	B	h max
USNM 501464	71·5	47·2	67	51	39	39	106	70	—	93
USNM 501343	66	42	59	46	28	38	112	66	—	80
USNM 501281b	58	37·5	56	38·2	22·5	25·5	117	72	—	70
USNM 501354	60	40	53·2	41·5	17	—	110	63	—	105
USNM 501347	40·7	26	36·5	30	15·2	—	110	64	—	44·5

Description. Moderate to large-sized for the genus, prosocline, inequilateral, equivalve. Valves flat or slightly inflated, with maximum inflation dorsocentral, moderately oblique. Valve outline subrectangular, elongated parallel to growth axis. Beak pointed, projecting above hinge line. Posterior auricle moderately large to large, well separated from disc. Growth axis straight to slightly convex anteriorly. Anterior margin relatively short, straight, passing into long anteroventral margin. Ventral margin narrowly rounded. Posterior margin concave at sulcus of posterior auricle.

Valves ornamented with distinct, sharp-edged, widely and more or less irregularly spaced concentric rugae. Inter-rugae spaces flat-floored, increase in width ventralward. Umbonal part usually with growth lines only; indistinct rugae may be developed in some individuals. Ornament passes onto posterior auricle, with rugae and growth lines curved markedly outward on posterodorsal part of auricle.

Remarks. Mytiloides scupini (Heinz, 1930) corresponds to forms previously referred to *Inoceramus frechi* in the sense of Andert (1911) (see Walaszczyk and Tröger 1996 for description and discussion). Small morphotypes with more regularly spaced rugae (Pl. 16, fig. 4) resemble *Mytiloides labiatoidiformis* (Tröger, 1967, p. 125, pl. 10, figs 5–6). *M. scupini* may be distinguished by its distinctly subtriangular and outwardly curved posterior auricle, which is usually small and narrow in *M. labiatoidiformis*. Adult fragments of *M. scupini* are very similar to those of *M. ratonensis* sp. nov., and may be inseparable.

One of the specimens illustrated by Scott and Cobban as *Inoceramus* aff. *perplexus* (1964, pl. 2, fig. 1), is referred to *M. scupini*. North American representatives of *M. scupini* were previously referred to *M. lusatiae*, *M. aviculoides*, *M. labiatoidiformis* or *M. kleini* (see synonymy). All these taxa are, however, quite distinct.

Inoceramus lusatiae Andert (1911, p. 54, pl. 2, fig. 1; pl. 3, fig. 3; pl. 8, figs 3–5), a member of the *lamarcki* group, possesses strong, round-topped rugae, with superimposed evenly-spaced growth lines; the widely extended, triangular, posterior auricle is smooth (see Walaszczyk and Wood 1999*a* for discussion

---

EXPLANATION OF PLATE 16

Figs 1–8, 10–11. *Mytiloides scupini* (Heinz, 1930). 1, USNM 501343; 2, USNM 501344; 3, USNM 501605; 4, USNM 501345; 5, USNM 501346; 6, USNM 501347; 7, USNM 501348; 8, USNM 501349; 10, USNM 501351; 11, USNM 501352: all USGS Mesozoic locality D12522 (Text-fig. 11, loc. 78)

Fig. 9. *Mytiloides ratonensis* sp. nov., USNM 501353, USGS Mesozoic locality D12522 (Text-fig. 11, loc. 78).

All ×1.

WALASZCZYK and COBBAN, *Mytiloides*

and illustration of paralectotypes from Andert's collection). We have not recognised *I. lusatiae* in the US Western Interior.

Mytiloides aviculoides is a basal Middle Turonian homeomorph of *M. scupini*, distinguished by its outwardly extended posterior auricle.

Mytiloides labiatoidiformis (Tröger, 1967, p. 125, pl. 10, figs 5–6) is a regularly ornamented species with slightly lamellate rugae and a markedly less extended posterior auricle. USNM 501345 (Pl. 16, fig. 4) closely resembles *M. labiatoidiformis*; we refer it to *M. scupini* on the basis of its relatively large posterior auricle. The concept of *M. labiatoidiformis* was considerably enlarged by Keller (1982), who included in the species forms that differ markedly from Tröger's original concept; none of the forms illustrated by him (varieties a and b) is conspecific with *M. labiatoidiformis*. Keller's *I. labiatoidiformis* var. b is well represented in North America and is conspecific with *Mytiloides ratonensis* sp. nov. (see Pl. 15, and illustration of the holotype in Scott *et al.* 1986, fig. 6h).

Inoceramus kleini Müller (1888, p. 415, pl. 18, fig. 1) differs in general outline, ornament and stratigraphical position. It was described originally from the lower Middle Coniacian [it co-occurs with *Volviceramus koeneni* (Müller, 1888) of the Subhercynian Cretaceous, Germany (Müller 1888, Assmus 1963)], and the typical form is known only in this interval. *I. kleini* is moderately to very inflated, slightly inequivalve with a well-separated but relatively small posterior auricle, and with very regular ornament of evenly spaced sharp-edged rugae. The central part of the disc bears much less regular, discontinuous radial ornament (see e.g. Walaszczyk 1992, pl. 37, fig. 3). Small forms similar to *Inoceramus perplexus* Whitfield, 1877 (see above) that occur in the uppermost part of the Lower Coniacian (Andert 1911, pl. 1, fig. 7; pl. 2, figs 6–8; Andert 1934, pl. 4, figs 9–11; pl. 5, figs 1–2; Heine 1929, pl. 2, figs 10–11; pl. 3, figs 12–13) have been assigned to *I. kleini* but there are no detailed studies of the succession leading from these forms to Middle Coniacian *I. kleini*. That these small forms are conspecific with *I. kleini*, the oldest of which were reported from the topmost *crassus* Zone (see Čech and Švábenická 1992; Walaszczyk and Wood, *in* Niebuhr *et al.* 1999), seems acceptable.

Occurrence: Upper Upper Turonian (excluding its topmost part) of the US Western Interior and Europe (England, France, Germany, Spain, Poland, Czech Republic, Romania, Russia). Common in the basal part (below the 'barren interval') of the Upper Turonian *scupini* Zone; much rarer in its upper part.

Mytiloides striatoconcentricus (Gümbel, 1868)

Text-figure 17

1868	*Inoceramus striato-concentricus* Gümbel, p. 69, pl. 2, fig. 4.
1928*c*	*Inoceramus striato-concentricus* Gümbel; Heinz, p. 68, pl. 4, fig. 3.
1939	*Inoceramus striatoconcentricus* Gümbel; Dacqué, p. 209, pl. 17, fig. 5.
non 1959	*Inoceramus striatoconcentricus* Gümbel; Dobrov and Pavlova, p. 135, pl. 2, figs 1–2 [pl. 2, fig. 1 = *Mytiloides* ex gr. *labiatus* (Schlotheim, 1813); pl. 2, fig. 2 = *Cremnoceramus waltersdorfensis hannovrensis* (Heinz, 1928)].
1967	*Inoceramus striatoconcentricus striatoconcentricus* Gümbel; Tröger, p. 84, pl. 9, figs 11–15, 17 (with synonymy).
?1971	*Inoceramus striato-concentricus* (Gümbel); Pergament, p. 59, pl. 8, fig. 2.
non 1971	*Inoceramus* cf. *striato-concentricus* (Gümbel); Pergament, p. 59, pl. 8, fig. 3.

EXPLANATION OF PLATE 17

Figs 1–8. *Mytiloides scupini* (Heinz, 1930). 1, USNM 501465; 2, USNM 501466; 3, USNM 501470; 4, USNM 501467; 5, USNM 501469; 6, USNM 501468; 8, USNM 501471: all USGS Mesozoic locality D10458 (Text-fig. 11, loc. 44). 7, USNM 501464, USGS Mesozoic locality D10459 (Text-fig. 11, loc. 45). 9, USNM 501346, USGS Mesozoic locality D12522 (Text-fig. 11, loc. 78). All ×1.

WALASZCZYK and COBBAN, *Mytiloides*

1982 *Inoceramus striatoconcentricus striatoconcentricus* Gümbel; Keller, p. 105, pl. 7, fig. 4.
1992 *Mytiloides striatoconcentricus* (Gümbel); Walaszczyk, p. 24 (*pars*), pl. 13, figs 4, 6.
non 1997 *Mytiloides striatoconcentricus* (Gümbel); Leckie *et al.*, fig. 36H–L.

Type. The neotype, designated by Dacqué (1939, p. 209, pl. 17, fig. 5), is the specimen from the Upper Turonian (?Lower Coniacian) of Grossbergsschichten, Germany, in Gümbel's collection.

Material. Five specimens from the lower Upper Turonian of the Belle Fourche section in the north-eastern flank of the Black Hills; four from USGS locality 27 (21192), and one from locality 28 (21194).

Description. Small to medium-sized: largest specimen, from locality 28 (21194) is 80 mm in axial length. The other specimens are 65, 50 and 45 mm (incomplete) long. Valves inequilateral, prosocline, markedly oblique; with δ between 40 and 50 degrees, and distinctly anteriorly curved growth axis. LV and RV identical in valve outline and general shape, but differ more or less in inflation, with RV less inflated. B/h ratio in more inflated LV attains almost 40 per cent. Maximum inflation dorsocentral. Anterior margin straight or slightly convex, with steep and high anterior face. Anteroventral margin long, wide, convex, passing into narrowly rounded ventral margin. Posterior margin concave. Hinge line moderately long (<50 per cent of relative axial length), straight. Posterior auricle small, triangular, elongated, and well separated from the disc.

Valves ornamented with fine, uniform, equidistant, raised growth lines over almost entire valve length, with only weak ventralward size increase. Few, very weak, low, indistinct rugae present on adult part of shell. Growth lines curve distinctly outward on the posterior auricle.

Remarks. We follow Tröger's (1967) interpretation, which agrees well with that of Gümbel (1868, p. 69, pl. 2, fig. 4). Diagnostic features of this species are the very fine, uniform ornament of raised growth lines and prominent inflation. The neotype is slightly more coarsely ornamented judging from Dacqué's (1939, pl. 17, fig. 5) illustration.

Mytiloides striatoconcentricus differs from *M. bellefourchensis* sp. nov. in its larger l/h ratio in the juveniles, its more ovate outline, and in its finer ornament; it is also distinctly more inflated. The ornament is similar to that of *M. herbichi* (Atabekian, 1969) (=*Inoceramus labiatus* var. *regularis* of Simionescu, 1899, pl. 2, fig. 3; see discussion and illustration in Walaszczyk and Szász 1997, p. 774), described from the Upper Turonian of Romania, which, however, possesses a triangular, posteriorly elongated posterior auricle, and is an almost flat upright form with straight growth axis.

Occurrence. *Mytiloides striatoconcentricus* is a rare species in the *I. dakotensis* Zone (middle and upper part of the *Scaphites whitfieldi* ammonite Zone) of the lower Upper Turonian of South Dakota, US Western Interior. Known from Germany, France, the Czech Republic, Romania, Poland, and Russia in Europe; occurs in the lower Upper Turonian of Mangyshlak, western Kazakhstan.

Genus CREMNOCERAMUS Cox, 1969
[*non Cremnoceramus* Heinz, 1932*b* (*nomen nudum*)]

Type species. By original designation, *Inoceramus inconstans* Woods, 1912, from the Lower Coniacian (Cox 1969, p. N315).

Remarks. For a description and discussion of *Cremnoceramus* Cox, 1969, see Kauffman (*in* Herm *et al.* 1979), Walaszczyk (1992), and Crampton (1996). The recent review by Collom (1998) is an attempt to interpret *Cremnoceramus* Heinz, 1932b, which is a *nomen nudum*. This author tried to follow Heinz's concept of a taxon that is unavailable under the Rules of the International Code of Zoological Nomenclature (ICZN), and must be rejected.

Walaszczyk and Wood (1999*a*; see also Walaszczyk 1992) gave an evolutionary and taxonomic interpretation of the genus *Cremnoceramus* based on material from Europe. The ancestral stock of the *Cremnoceramus* clade is the *C. waltersdorfensis* lineage, which appeared in the latest Turonian and

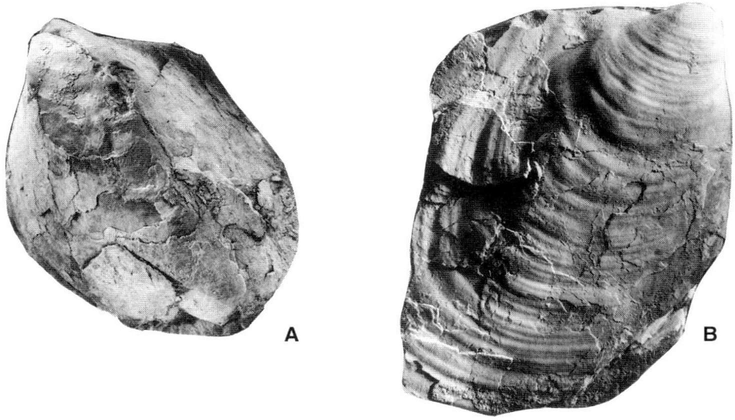

TEXT-FIG. 17. *Mytiloides striatoconcentricus* (Gümbel, 1868); North Belle Fourche section, South Dakota, USGS Mesozoic locality 21192 (Text-fig. 11, loc. 27; see also Text-fig. 6). A, USNM 501450; B, USNM 501451: both ×1.

ranged, most probably, to the end of the Early Coniacian. Two distinct chronosubspecies may be distinguished within this lineage: *Cremnoceramus waltersdorfensis waltersdorfensis* (Andert, 1911) and *C. waltersdorfensis hannovrensis* (Heinz, 1928). Two other lineages rapidly arose from this lineage by cladogenesis: the earlier *deformis* and later *crassus* lineages. The earliest member of the *deformis* lineage is *Cremnoceramus erectus*, the *Cremnoceramus rotundatus sensu* Tröger *non* Fiege of authors, and is referred accordingly to *C. deformis erectus* (Meek). It was succeeded by *C. deformis dobrogensis* (Szász, 1985) and *C. deformis deformis* (Meek, 1871). The later *C. waltersdorfensis hannovrensis* gave rise to the *Cremnoceramus crassus* lineage, composed of *C. crassus crassus* and *C. crassus inconstans* (Text-fig. 21). The appearance of *C. deformis erectus*, the earliest of the *deformis* lineage, marks the base of the Coniacian stage. This evolutionary sequence can also be recognised in the US Western Interior, even though the Lower Coniacian inoceramid record is dominated by the *deformis* lineage rather than the *waltersdorfensis* lineage, as in Europe.

Some specimens of *Cremnoceramus* attain a remarkable size in the adult stage (Text-figs 18–20). They have a normal-sized juvenile stage, but their adult stage is 1·5–2 times larger than in typical specimens. The forms illustrated have a juvenile stage identical to that in *C. deformis dobrogensis* (Szász). In the material from the Shelby section in bed K (*crassus* Zone), specimens with a similar, extraordinarily extended adult stage but with *crassus*-like juveniles also occur. Forms with extremely elongated adults were also noted in the European *waltersdorfensis* lineage, in the *waltersdorfensis* Event of Poland (Walaszczyk 1992, pl. 17, fig. 4), and large individuals with an *inconstans*-like juvenile stage were illustrated by Egoyan (1955, pl. 5, fig. 1) from the Caucasus. An abnormally elongated adult stage is also present in New Zealand [*Cremnoceramus bicorrugatus bicorrugatus* (Marwick, 1926); see Crampton 1996, p. 75, pl. 20, figs A–C].

This gigantism is here interpreted as phenotypic, and the result of ecological conditions. Large taxa based on this feature, namely *Inoceramus browni* Cragin, 1889, *I. gradatus* Egoyan, 1955, and *I. pseudoinconstans* Szász, 1985, are regarded as synonyms of *Cremnoceramus deformis dobrogensis*, *C. crassus inconstans*, and *C. crassus inconstans/crassus* respectively, on the basis of the characteristics of their juvenile stage.

Occurrence. Known from the uppermost Turonian and Lower Coniacian of Europe (England, France, Spain, Germany, Czech Republic, Romania, Bulgaria, Poland, Ukraine, Russia, Sweden), North America (Gulf Coast, Western Interior, Alaska, Pacific coast), Japan, Pacific Russia, New Zealand, South America (Brazil), and South Africa.

TEXT-FIG. 18. *Cremnoceramus deformis* (Meek); morphotype *browni* Cragin, USNM 501459, USGS Mesozoic locality D12426 (Text-fig. 11, loc. 21); ventral view; × 0·4.

Cremnoceramus waltersdorfensis (Andert) lineage

Remarks. Detailed discussion of this lineage was recently provided by Walaszczyk and Wood (1999a). The Western Interior record is much poorer than in Europe. The exception is the entrance level of the lineage, the *waltersdorfensis* Event, which is well represented in the Shelby section in northern Montana, and farther south in the Pueblo-La Junta sections in south-east Colorado, and the Springer section in New Mexico. Higher up in the Lower Coniacian succession, representatives of the lineage are sporadic. Despite this, the lineage demonstrated in Europe (Text-fig. 21) with *C. waltersdorfensis waltersdorfensis* and *C. waltersdorfensis hannovrensis* at the base of the *C. deformis dobrogensis* Zone, is recognisable in the Western Interior.

Occurrence. Uppermost Turonian–Lower Coniacian of North America (Western Interior, Gulf Coast, Alaska), Europe (England, France, Germany, Spain, Czech Republic, Romania, Poland, Ukraine, Russia), Asia (Kazakhstan, Turkmenistan, Afghanistan, Japan), South Africa, and South America (Brazil).

Cremnoceramus waltersdorfensis waltersdorfensis (Andert, 1911)

Plate 18, figures 1–9

1911 *Inoceramus waltersdorfensis* Andert, p. 53, pl. 5, figs 2, 5.

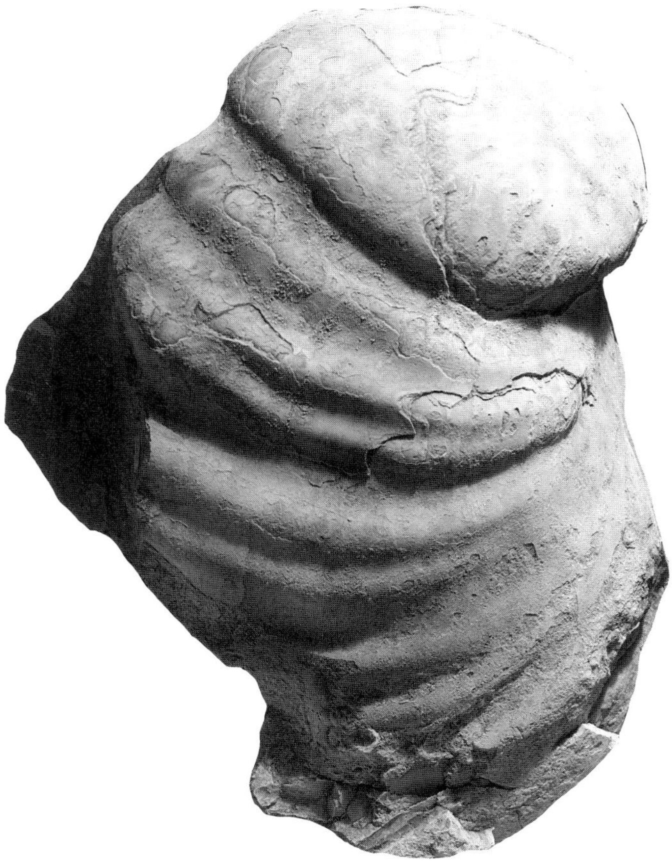

TEXT-FIG. 19. *Cremnoceramus deformis* (Meek); morphotype *browni* Cragin, USNM 501459, USGS Mesozoic locality D12426 (Text-fig. 11, loc. 21). Anterior view of specimen in Text-fig. 22; × 0·4.

1911 *Inoceramus sturmi* Andert, p. 58, fig. 5.
1934 *Inoceramus waltersdorfensis* Andert; Andert, p. 112, pl. 4, figs 2–7.
1967 *Inoceramus waltersdorfensis waltersdorfensis* Andert; Tröger, p. 114, pl. 12, figs 1–2; pl. 13, figs 1–5.
1967 *Inoceramus inconstans inconstans* Woods; Tröger, p. 101, pl. 13, fig. 19.
1977*b* *Inoceramus waltersdorfensis waltersdorfensis* Andert; Kauffman (*pars*), pl. 9, fig. 22 (?*non* pl. 9, fig. 26).
1978*a* *Inoceramus waltersdorfensis waltersdorfensis* Andert; Kauffman, pl. 5, figs 6–7.
1978*a* *Inoceramus waltersdorfensis hannovrensis* Heinz; Kauffman, pl. 5, figs 3, 15.
1979 *Cremnoceramus*? *waltersdorfensis hannovrensis* (Heinz); Kauffman, *in* Herm *et al*., p. 59, pl. 9, figs D, G.
1984 *Cremnoceramus*? cf. *waltersdorfensis* (Andert); Cobban, p. 8, fig. 1.
1989 *Cremnoceramus*? cf. *waltersdorfensis* (Andert); Kennedy *et al*., p. 111, fig. 34L.
1992 *Cremnoceramus waltersdorfensis* (Andert); Walaszczyk, p. 41 (*pars*), text-fig. 12, pl. 16, figs 1–11; pl. 17, figs 1–5; pl. 18, figs 1–3.
1992 *Inoceramus waltersdorfensis waltersdorfensis* Andert; Elder and Box, p. 25 (*pars*), figs 13.2–3 (?*non* fig. 1).
1996 *Cremnoceramus waltersdorfensis* (Andert); Walaszczyk, p. 374, figs 3C–F, 4F–G.
1997 *Inoceramus* aff. *ernsti* Heinz; Leckie *et al*. (*pars*), fig. 37C–D (?*non* fig. 37E–G).
1999*a* *Cremnoceramus waltersdorfensis waltersdorfensis* (Andert); Walaszczyk and Wood, p. 413, pl. 5, figs 1, 3–7, 9–13, 15–18; pl. 15, figs 1–3; pl. 17, fig. 3 (with full synonymy).

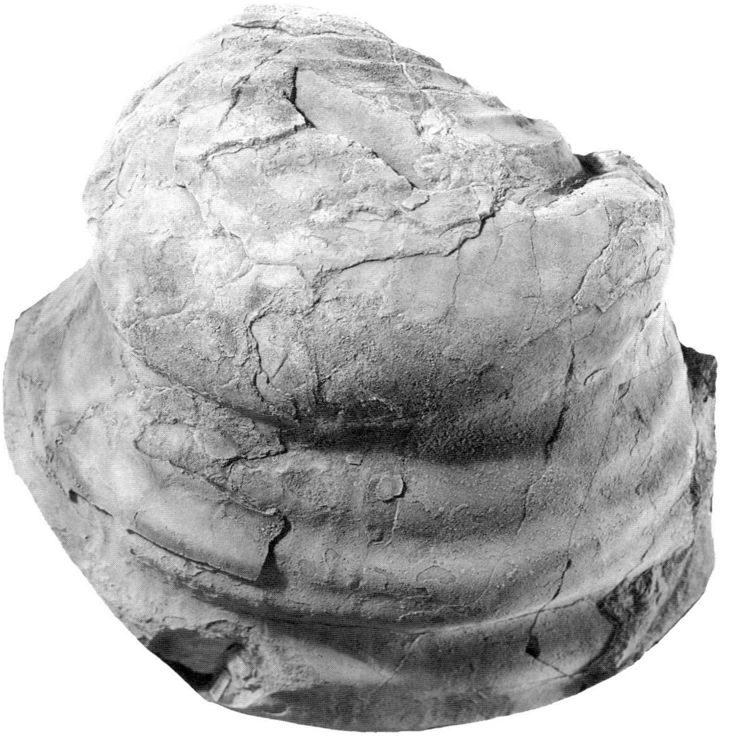

TEXT-FIG. 20. *Cremnoceramus deformis* (Meek); morphotype *browni* Cragin, USNM 501460, USGS Mesozoic locality D12426 (Text-fig. 11, loc. 21); × 0·4.

Types. Lectotype by subsequent designation of Walaszczyk (1992, p. 41) is the original of Andert (1911, pl. 5, fig. 5, and re-illustrated by Walaszczyk 1996, fig. 3E–F). Paralectotype is Andert's original (1911, pl. 5, fig. 2; re-illustrated by Walaszczyk 1996, fig. 3C–D); both housed in the State Museum of Mineralogy and Geology in Dresden, Germany.

Material. Numerous specimens from the Springer, Pueblo and La Junta sections (from the *waltersdorfensis* Event), and also from the same event in the Shelby and Johnson Bridge sections.

Dimensions (*mm*).	h	l	H	L	s	VR	α	δ	h max
USNM 501357	44·5	40·5	40	39	28	27	126	73	49
USNM 501363	29	31·5	29	31·5	27	27	—	80	47
USNM 501360	42·5	39	42	39	26·5	23·5	121	76	61
USNM 501362	35	27	35	29	21	16·5	118	78	46
USNM 501379a	40·5	39	41	37	25	27	116	80	45
USNM 501368	42	38	40·2	39	29	24·5	118	78	45·5
USNM 501369	36	33·8	34·8	33·5	25	22	112	72	46·5
USNM 501370	40	34	38·1	33·9	23	26	105	75	55
USNM 501373	32·5	29	30·3	28	20·5	17	114	70	34
USNM 501374	30·4	31	—	—	21·5	18	120	70	36·5
USNM 501375	39	37	39	37	—	18	118	75	64
USNM 501376	46·2	42·5	45	—	—	32	119	90	52
USNM 501377	45	37	44·8	35·5	22	29	120	80	51·5
USNM 501378	42	35	39·5	36·5	23	26	112	79	52·5
USNM 501372	38	34	35	36·8	25·5	18	120	69	47
USNM 501371	37	33	36	34	22	18	118	75	64

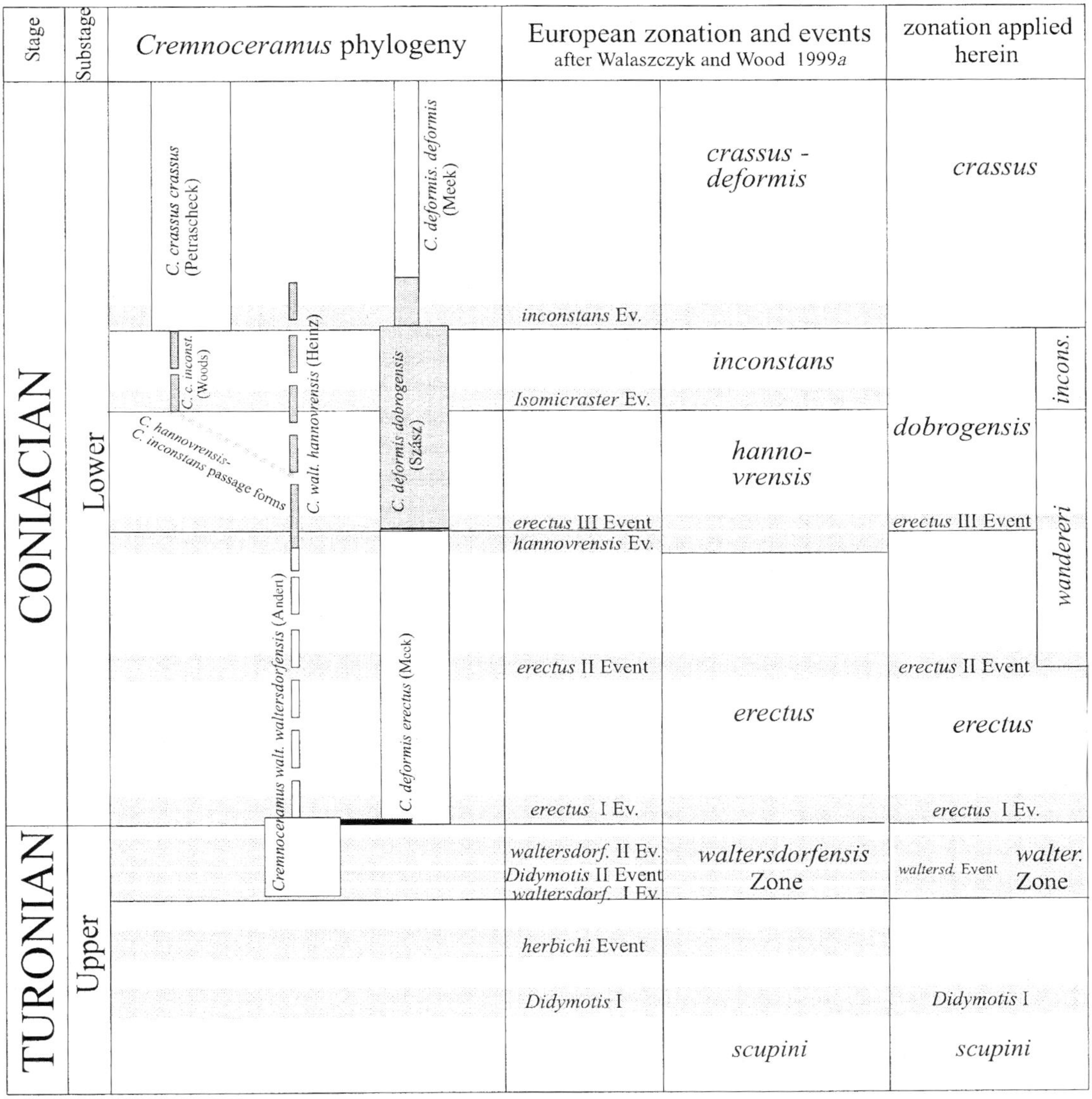

TEXT-FIG. 21. Phylogeny of *Cremnoceramus* Cox, 1969; modified after Walaszczyk and Wood (1999*a*).

Description. Small to moderate-sized, inequilateral, equivalve. Outline subquadrate to subrounded, slightly prosocline to slightly orthocline. Juvenile stage slightly to moderately inflated. Anterior margin long, straight, passing into broadly rounded ventral margin. Posterior margin concave. Hinge line long, straight. Posterior auricle large, extended posteriorly, either poorly delimited or well separated from disc along well-developed auricular sulcus. Anterior face steep, posterior face flattened.

Juvenile ornament composed of raised, slightly asymmetrical, somewhat lamellate growth lines and more or less irregular rugae. Rugae may be very weakly developed in some specimens (Pl. 18, figs 3, 6). Adult stage smooth, or with irregular rugae.

Remarks. See Walaszczyk (1996) and Walaszczyk and Wood (1999*a*) for recent description, discussion, and illustration of the European specimens, including the types.

C. waltersdorfensis waltersdorfensis is relatively variable in respect of both general outline and surface ornament. It is succeeded by *C. waltersdorfensis hannovrensis* within the *C. deformis dobrogensis* Zone in both Europe and the USA, but is extremely rare in the Western Interior. Earlier reports of the subspecies *hannovrensis* from the Turonian/Coniacian boundary interval refer to the nominate subspecies.

Of the specimens from Alaska illustrated and referred to *I. waltersdorfensis waltersdorfensis* by Elder and Box (1992), two (Elder and Box 1992, fig. 13.2–3) are most probably conspecific. Their third specimen (1992, fig. 13.1) has an umbo raised distinctly above the hinge line, is more slender, and has a different ornament. It represents some other Upper Turonian species.

Occurrence. Topmost Turonian (*waltersdorfensis* Zone) through lowermost Coniacian (*erectus* Zone) of Europe (England, Germany, France, Spain, Poland, Czech Republic, Romania, Ukraine, Russia) and North America (Western Interior, Gulf Coast, Alaska). Some of the specimens illustrated by Matsumoto and Noda (1985), and referred by them to *Inoceramus rotundatus* (Fiege), are very similar to *C. waltersdorfensis waltersdorfensis*. It is thus possible that the species ranges to the North Pacific Province. In the Western Interior it appears in flood abundance in the *waltersdorfensis* Event, and becomes extremely rare higher in the Lower Coniacian succession.

Cremnoceramus waltersdorfensis hannovrensis (Heinz, 1932)

Plate 26, figure 2

1932*a*	*Inoceramus hannovrensis* Heinz, p. 29.
1959	*Inoceramus striato-concentricus* Gümbel; Dobrov and Pavlova, p. 135, pl. 2, fig. 2.
1967	*Inoceramus waltersdorfensis hannovrensis* Heinz; Tröger, p. 117, pl. 12, figs 3–4; pl. 13, figs 6–9.
1968	*Inoceramus waltersdorfensis* Andert; Kotsubinsky, p. 127, pl. 18, figs 2, ?3.
1982	*Inoceramus waltersdorfensis hannovrensis* Heinz; Keller, p. 112, pl. 8, fig. 3.
1989	*Inoceramus* (?*Cr.*) *waltersdorfensis hannovrensis* Heinz; Küchler and Ernst, pl. 4, figs 2–4.
1992	*Cremnoceramus waltersdorfensis* (Andert); Walaszczyk, p. 41, pl. 18, figs 4–7; pl. 19; pl. 30, fig. 1[(*non* pl. 16, figs 1–11; pl. 17, figs 1–5; pl. 18, figs 1–3 = *Cremnoceramus waltersdorfensis waltersdorfensis* (Andert, 1911)]
non 1992	*Inoceramus* cf. *I. waltersdorfensis hannovrensis* Heinz; Elder and Box, p. 26, fig. 13.4–7.
1998	*Cremnoceramus waltersdorfensis hannovrensis* (Heinz); Walaszczyk and Wood, pl. 5, figs 2, 8; pl. 9, fig. 2; pl. 11, fig. 2; pl. 12, figs 1–4; pl. 13, figs 2–7; pl. 15, figs 4–5 (with synonymy).
non 1998	*Cremnoceramus waltersdorfensis hannovrensis*; Collom, fig. 4A [= *Cremnoceramus deformis erectus* (Meek)].

Type. The lectotype, designated by Tröger (1967, pl. 12, fig. 3), is no. 914, housed in the collections of Museum of the Niedersächsiches Landesamt für Bodenforschung, Hannover, Germany. Heinz' (1932*a*, p. 29) description is a *nomen nudum* that was validated subsequently by Tröger (1967).

Material. Rare specimens from the Pueblo area and the Shelby section.

EXPLANATION OF PLATE 18

Figs 1–9. *Cremnoceramus waltersdorfensis waltersdorfensis* (Andert, 1911). 1, USNM 501355; 2, USNM 501356: both USGS Mesozoic locality D1781 (Text-fig. 11, loc. 61). 3, USNM 501357, USGS Mesozoic locality 21421 (Text-fig. 11, loc. 16). 4, USNM 501358, USGS Mesozoic locality 21421 (Text-fig. 11, loc. 16). 5, USNM 501359, USGS Mesozoic locality 21421 (Text-fig. 11, loc. 16). 6, USNM 501360, USGS Mesozoic locality D11939 (Text-fig. 11, loc. 43). 7, USNM 501361, USGS Mesozoic locality D1781 (Text-fig. 11, loc. 61). 8, USNM 501362; 9, USNM 501363: both USGS Mesozoic locality D1781 (Text-fig. 11, loc. 61).

Figs 10–13. *Mytiloides scupini* (Heinz, 1930). 10, USNM 501364, USGS Mesozoic locality D3681 (Text-fig. 11, loc. 74). 11, USNM 501365, USGS Mesozoic locality D3681 (Text-fig. 11, loc. 74). 12, USNM 501367, USGS Mesozoic locality D8842 (Text-fig. 11, loc. 39). 13, USNM 501366, USGS Mesozoic locality D12522 (Text-fig. 11, loc. 78).

All × 1.

WALASZCZYK and COBBAN, *Cremnoceramus, Mytiloides*

Remarks. We follow Walaszczyk and Wood (1999*a*) and interpret *hannovrensis* as a successor of *C. waltersdorfensis waltersdorfensis* (Andert, 1911). *C. walt. hannovrensis* differs from *C. walt. waltersdorfensis* in being more prosocline, larger in average size, with a long, straight and steep anterior face and well-developed umbonal stage with the beak projected well above the hinge line.

The specimen illustrated by Collom (1998, fig. 4A) differs from representatives of the *waltersdorfensis* lineage in its ornament of evenly spaced, slightly lamellaete rugae, typical for young and/or early representatives of *C. deformis erectus* (Meek, 1877).

The specimens from Alaska, compared with subspecies *hannovrensis* by Elder and Box (1992, fig. 13.4–7), differ in the development of the umbonal region, anterior margin, and ornament. They should be referred to *Inoceramus kuskowimensis* Elder and Box, 1992.

Occurrence. Beginning with the *hannovrensis* Event, the subspecies *hannovrensis* (Heinz) is abundant in Europe (Germany, France, Spain, England, Poland, Czech Republic, Romania, Ukraine, Russia) and western Asia (Kazakhstan, Turkmenistan), and dominates the inoceramid fauna up to the appearance level of *C. inconstans* (see Walaszczyk and Wood 1999*a*). In the Western Interior, Heinz's form is very rare, and inoceramid assemblages are dominated by representatives of the *deformis* lineage.

Cremnoceramus deformis (Meek, 1871) lineage

Following Walaszczyk and Wood (1999*a*), this lineage is interpreted as comprising upright, weakly to moderately oblique, rugate cremnoceramids. It arose at the beginning of the Coniacian (see Text-fig. 21) where its appearance marks the stage boundary and ranges throughout the substage. More than 300 Western Interior specimens were studied, including the types of *C. erecus*, *C. deformis*, and numerous topotype specimens. These, and the European specimens of the lineage (Szász 1985; Walaszczyk 1992, 1996; Walaszczyk and Wood 1999*a*), reveal the following:

1. The lineage underwent phyletic change in mean adult size, change in spacing and distribution of rugae and decrease in l/h ratio (see Text-figs 22–23).

2. The rate of change of characters was apparently slow during most of the Early Coniacian but punctuated by two intervals of abrupt, step-wise change, as a result of which it is possible to distinguish three distinct intervals each characterized by a distinct chronosubspecies. The levels of abrupt change are well constrained in Western Interior sections, at the base of the *dobrogensis* Zone and higher, within the *crassus* Zone. The part of the lineage that succeeded the *erectus* interval is very poorly represented in Europe, but the two levels appear to correlate with the base of the *waltersdorfensis hannovrensis* Zone and a lower part of the *crassus/deformis* Zone (probably the level of the *deformis* Event: see Ernst *et al.* 1983). The chronosubspecies are, in ascending order:

Cremnoceramus deformis erectus (Meek, 1877): small to medium size, l/h ratio < 1, concentric rugae closely and regularly spaced (see Text-figs 22–23). Meek's lectotype falls among very early representatives of the subspecies on the basis of ornament characters.

C. deformis dobrogensis (Szász, 1985): moderate to large size, l/h ratio $\cong 1$ or > 1: subquadrate or lower than long, with rugae evenly spaced in the juvenile, widely spaced thereafter.

C. deformis deformis (Meek, 1871): size and l/h characteristics as in subspecies *dobrogensis*, but with rugae becoming progressively wider spaced through ontogeny (see Text-figs 22–23).

3. The obliquity, character of the umbonal region, posterior auricle (size and distinctness from disc), and anterior margin, show approximately the same range of variation within any population of the lineage.

The *deformis* lineage, following Cobban (1951*b*), Scott and Cobban (1964), Kauffman (1977*b*; *in* Kauffman *et al.* 1978, 1993; *in* Herm *et al.* 1979), Ernst *et al.* (1983) and Wood *et al.* (1984), has been widely used for Lower Coniacian zonation, with the members of the lineage providing indices for three successive *rotundatus*, *erectus*, and *deformis* biozones. The concept of particular species within the lineage, as interpreted by Kauffman in North America and by European workers, is markedly different. Both *rotundatus* and *erectus* as interpreted by Kauffman (e.g. *in* Kauffman *et al.* 1993) correspond to *rotundatus* as used in Europe (e.g. Wood *et al.* 1984). *Cremnoceramus erectus* as used by Wood *et al.*

(1984) falls into a range of *deformis* as used by Kauffman. Kauffman (1977*b*; *in* Kauffman *et al.* 1978, 1993; *in* Herm *et al.* 1979) used the name *rotundatus* (*sensu* Tröger), for small, early representatives of the lineage, but even in the basal *erectus* I Event (*rotundatus* Event of former usage), they are already associated with forms close to the lectotype of *C. erectus* (Meek).

Above the *deformis* Zone in the Western Interior, the inoceramid record is dominated by *C. crassus crassus* (Petrascheck, 1903) of the *crassus* lineage. This lineage evolved from the *waltersdorfensis* lineage at the base of the *inconstans* Zone (Walaszczyk and Wood 1999*a*; Text-fig. 21 herein), a level that corresponds to the middle part of the *deformis dobrogensis* Zone, as indicated by the presence of rare specimens of *C. inconstans* from the same interval. Thus the succession *erectus–deformis–crassus* that dominates the Western Interior inoceramid record, is not a record of the evolution of a single lineage, as proposed by Kauffman (*in* Kauffman *et al.* 1978, 1993), but rather of changes in dominance of two distinct lineages in the record.

Cremnoceramus deformis erectus (Meek, 1877)

Plates 19–24; Plate 25, figures 1–2, 5–8

1877 *Inoceramus erectus* Meek, p. 145, pl. 13, figs 1, 1a; pl. 14, fig. 3.
1894 *Inoceramus deformis* Meek; Stanton, p. 85, pl. 14, fig. 1; pl. 15, fig. 1 (only).
1898 *Inoceramus deformis* Meek; Logan, p. 486, pl. 96, fig. 1 (only).
1912 *Inoceramus inconstans* Woods, p. 285, text-fig. 44.
1964 *Inoceramus erectus* Meek; Scott and Cobban, pl. 2, fig. 6.
1967 *Inoceramus rotundatus* Fiege; Tröger, p. 110, pl. 12, figs 5–6; pl. 13, figs 10–13.
1967 *Inoceramus inconstans lueckendorfensis* Tröger, p. 102, pl. 11, figs 1–2.
1975 *Inoceramus erectus* Meek; Hattin, pl. 2, fig. 12.
1977*b* *Inoceramus erectus erectus* Meek; Kauffman, pl. 11, figs 3–4.
1977*b* *Inoceramus erectus* Meek, n. ssp. ('late form'); Kauffman, pl. 11, fig. 6.
1978 *Inoceramus erectus erectus* Meek; Kauffman *et al.*, pl. 15, figs 3–4.
1978 *Inoceramus erectus* Meek, n. ssp. ('late form'); Kauffman *et al.*, pl. 15, fig. 6.
1979 *Cremnoceramus*? sp. aff. *C.*? *rotundatus* (Fiege); Kauffman, *in* Herm *et al.* 1979, p. 68, pl. 9A.
1979 *Inoceramus* sp. aff. *I. ernsti* Heinz; Kauffman, *in* Herm *et al.* 1979, pl. 9B.
1979 *Cremnoceramus*? *rotundatus* (Fiege); Kauffman, *in* Herm *et al.* 1979, p. 68, pl. 9C.
1982 *Inoceramus rotundatus* Fiege; Keller, p. 114, pl. 8, fig. 2.
1984*b* *Cremnoceramus*? *rotundatus* (Fiege); Cobban, p. 9, pl. 1, figs 2–3.
1985 *Inoceramus rotundatus* Fiege; Szász, p. 167, pl. 10, fig. 4; pl. 30, fig. 7.
non 1985 *Inoceramus erectus* Meek; Szász, p. 162, pl. 1, fig. 2.
1989 *Cremnoceramus* cf. *rotundatus* (Fiege); Kennedy *et al.*, p. 111, fig. 34J–K.
1992 *Inoceramus* (*Cremnoceramus*?) sp. with affinities to *I.* (*C.*?) *rotundatus* Fiege and *I* (*C.*?) *erectus* (Meek); Elder and Box, p. 26, fig. 16.5.
1992 *Cremnoceramus brongniarti* (Mantell); Walaszczyk, p. 48, pl. 22, figs 1–3; pl. 23, figs 1–5; pl. 24, figs 1–5; pl. 25, figs 1–5; pl. 30, fig. 2.
1996 *Cremnoceramus rotundatus* (Tröger 1967 *non* Fiege 1930); Walaszczyk, p. 377, fig. 3G–J.
1998 Transitional form between *C. tarlovensis* and *C. erectus*; Collom, fig. 2D.
1998 Transitional form between *C. waltersdorfensis* complex and *C. tarlovensis*; Collom, fig. 2E.
1998 *Cremnoceramus tarlovensis*; Collom, figs 2F–G, 4B–C.
1998 *Cremnoceramus erectus*; Collom, fig. 4D.
1998 *Cremnoceramus waltersdorfensis hannovrensis*; Collom, fig. 4A.
1998 *Cremnoceramus deformis*; Collom, fig. 6C–D.
1999*a* *Cremnoceramus deformis erectus* (Meek); Walaszczyk and Wood, p. 415, pl. 5, fig. 14; pl. 6, figs 1–6, 8; pls 7–8; pl. 9, figs 1, 3–6; pl. 10, figs 1–4, 6; pl. 11, figs 1, 3, 5–7; pl. 12, fig. 1; pl. 13, fig. 1; pl. 15, fig. 6).

Types. The lectotype, here designated, is USNM 7850, the double-valved specimen illustrated by Meek (1877, pl. 13, fig. 1; Pl. 19, fig. 1 herein). Meek's second specimen (1877, pl. 13, fig. 1a; Pl. 19, fig. 3 herein), and one unillustrated specimen are paralectotypes.

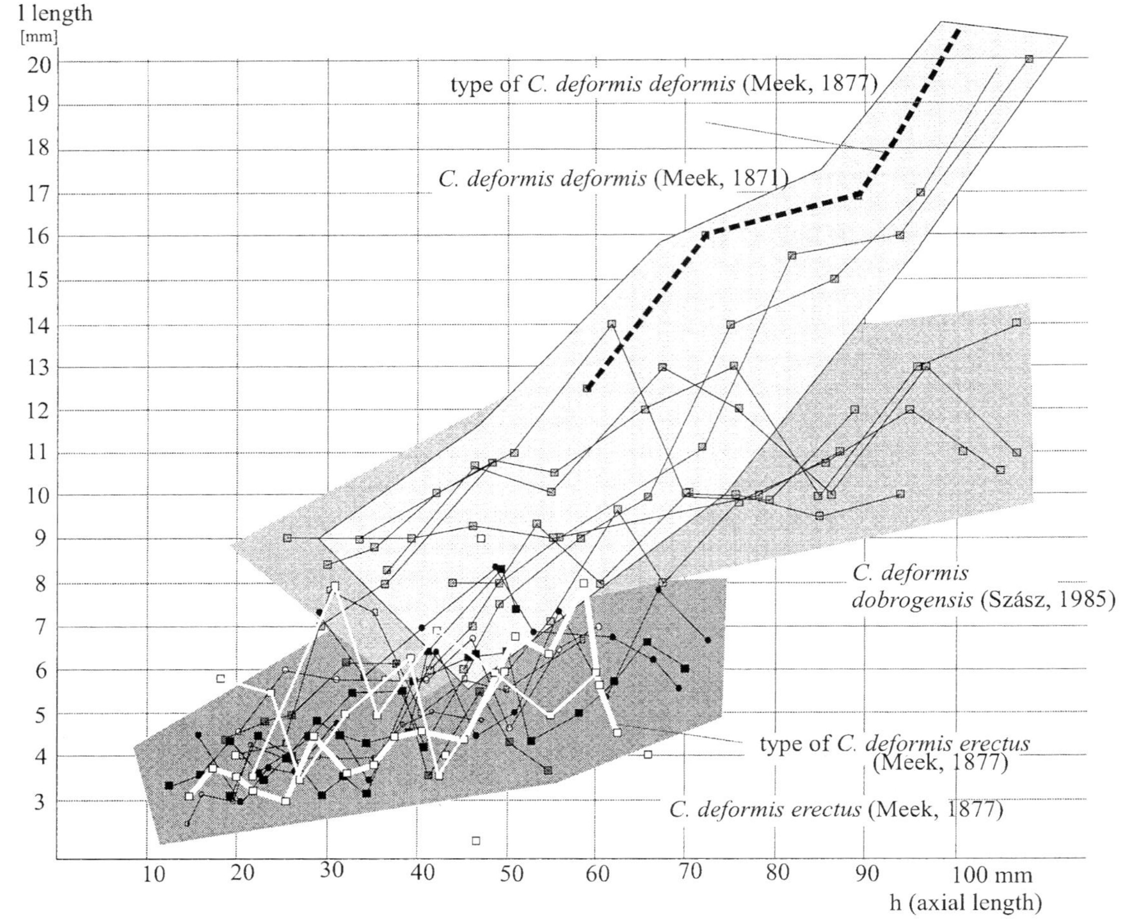

TEXT-FIG. 22. Diagram showing the distribution of rib spacing within the *deformis* lineage; types of *C. erectus* and *C. deformis* are shown in each case.

Meek (1877) indicated Chalk Creek, Uptown, Utah, as the type locality of his new species. Upton (as it is now spelt) is a small village about 15 km east of Coalville (Text-fig. 24), to the east of Salt Lake City, northern Utah. The only inoceramid-bearing beds found near this locality are light-yellow, fine-grained, massive sandstones in the Grass Creek Member of the upper part of the Frontier Formation (see Text-fig. 24, and Ryer 1976, 1977*a*, *b*). This unit, referred by Trexler (1966) to a separate Meadow Creek Member, was subsequently reinterpreted by Hale (1960, 1976) and Ryer (1976, 1977*a*, *b*) and referred to an upper part of the Grass Creek Member, a lithostratigraphical unit well represented in the environs of Coalville. Along Huff Creek, 4·5 km north of Upton, where the creek forks (Text-fig. 24), marine near-shore sandstones crop out as a pair of ridge-forming beds. The more westerly bed has a rich inoceramid fauna. *Cremnoceramus deformis erectus* (Meek, 1877) occurs in the lower third of this. Fossils seem to be absent from the middle, less resistant part but appear again in the upper ridge-forming unit, where inoceramids are represented by *Cremnoceramus deformis dobrogensis* (Szász, 1985). Inoceramids, mostly double-valved, occur here in patches several metres wide.

Material. Numerous specimens from much of the Western Interior.

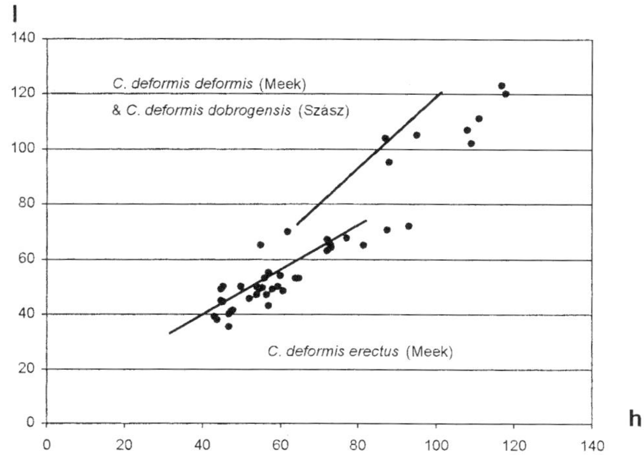

TEXT-FIG. 23. h/l ratio in the *deformis* lineage.

Dimensions (mm).	h	l	H	L	s	VR	α	δ	h max
USNM 501403	73	68·5	73	68	—	50	115	74	—
USNM 501381	76	67	74	65·5	40	44	114	78	97
USNM 501396	74	63	73	65	39	51	106	67	88
USNM 501389	92	71	92	75	47	60	120	80	—
USNM 501411	77	63	70	67	44	41	110	65	87
USNM 501412	71·5	58	68	58	42	46	110	75	78
USNM 501399	62	46	58·5	50·5	34	36	108	69	65
USNM 501401	44	37	40	40	29	27	106	65	53
USNM 501400	50	46	49	46·5	33	33	110	70	62
USNM 501406	54	51	52	51	34	37	115	74	65
USNM 501405	55	47	51	46	36	34	106	72	—
USNM 501404	60	50	58	54	45	42	122	76	71
USNM 501397	74	63	70	64	47	42	110	69	88
USNM 501402	63	58	60	54	39	38	115	70	76

Description. The lectotype (Pl. 19, fig. 1) is a double-valve internal mould in yellow sandstone, with traces of the outer prismatic layer preserved. Valves inequilateral, with subquadrate outline, inequivalved in respect of measured adult axial length owing to slight post-mortem dorso-ventral compression of RV, which is consequently more inflated than LV (B of RV and LV 32 and 36 mm respectively). The length of both valves measured from beak to venter at the lateral surface along the growth axis equal in both valves. Anterior margin long, slightly convex (almost straight), passing into the rounded ventral margin. Posterior margin too poorly preserved for description. Beak projects slightly above long, straight hinge line. Anterior face steep, uniformly inflated, without distinct geniculation. Boundary between juvenile and adult stages clearly visible at 58·6 and 54·3 mm axial length in LV and RV respectively, owing to the abrupt ornament change: regular rugae replaced by irregular rugae, or it is smooth.

Paralectotype (Pl. 19, fig. 3) is a single LV, which is markedly compressed laterally. Small portion only of the adult stage is preserved. Umbo projects markedly above hinge line (partly as a result of lateral compression). Anterior margin slightly convex, very long, passing into rounded ventral margin. Posterior margin straight. Hinge line long, straight. Posterior auricle prominent, well separated from the disc. Ornament almost completely effaced, with only small parts of two rugae preserved at the juvenile/adult junction.

Small to medium-sized, inequilateral, equi- to subequivalved. Outline subquadrate to subrounded, orthocline to slightly prosocline. Anterior margin straight or slightly convex; may be slightly concave near umbo. Anterior face steep. Ventral margin broadly rounded, passing into straight or convex posterior margin. Hinge line moderately long to very long. Growth axis straight, may be slightly convex anteriorly. Umbonal part moderately or weakly separated from posterior auricle. Beak pointed, curved antero-dorsally. Posterior auricle variable; in more upright forms it is large,

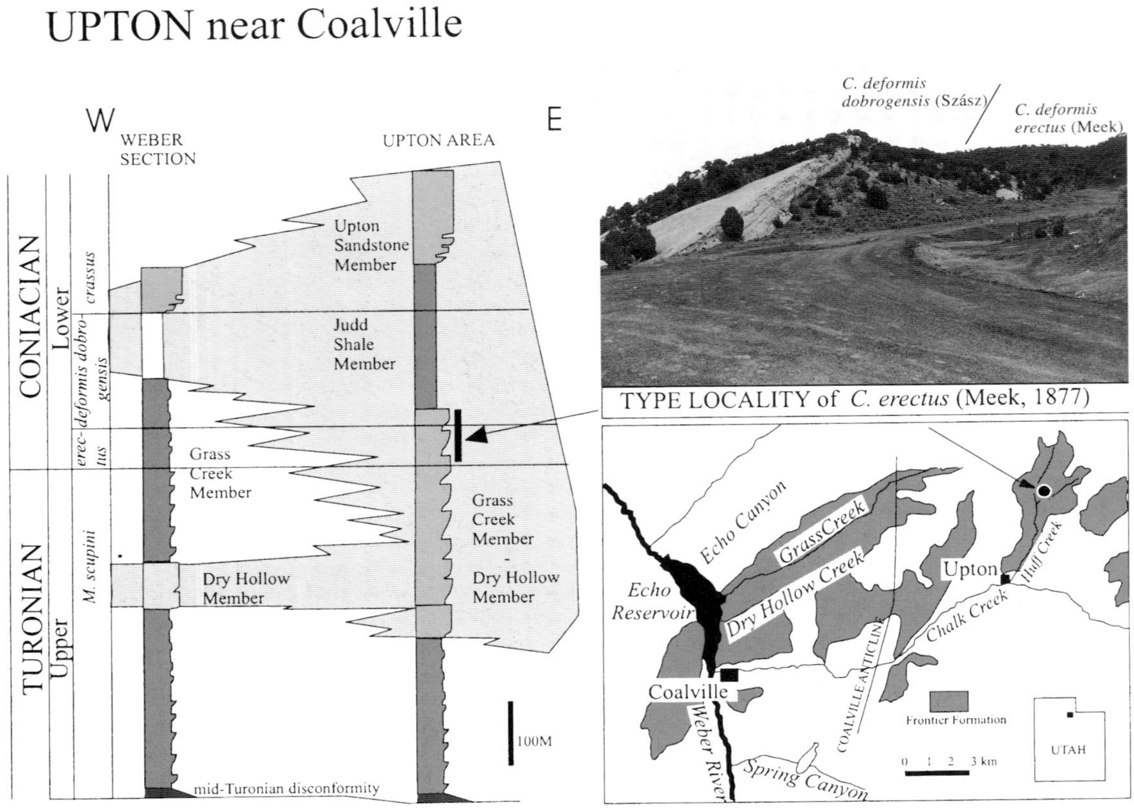

TEXT-FIG. 24. Geographical and stratigraphical location of the type locality of *Cremnoceramus erectus* (Meek, 1877); map and sections from Ryer (1977*a*, *b*).

subtriangular, extended posteriorly, well separated from disc; usually less extended and less distinctly separated from disc in prosocline forms. Juvenile ornament composed of subeven, closely spaced, concentric rugae, sometimes with intercalated ribs. Umbonal part up to 15–20 mm axial length, covered exclusively by raised, lamellate growth lines. Adult with irregular rugae or completely smooth.

Ligament thin to moderately thick, developed evenly in both valves. Ligamental plate varies from very low (< 1 mm) to moderately high (up to 4·6 mm in measured specimens). Shape of resilifers, density, and relations to inter-resilifer areas markedly variable.

Remarks. *Cremnoceramus deformis erectus* varies considerably in surface ornament, spacing and character of rugae, which range from fine and closely spaced (Pl. 22, figs 1, 4–5; Pl. 25, fig. 7) to

---

EXPLANATION OF PLATE 19

Figs 1–3. *Cremnoceramus deformis erectus* (Meek, 1877). 1, USNM 7850, plaster cast of the lectotype (original of Meek, 1877, pl. 13, fig. 1): a, lateral view of the RV; b, anterior view; c, lateral view of the LV. 2, USNM 501380, USGS Mesozoic locality D8284 (Text-fig. 11, loc. 80): a, lateral view of the RV; b, anterior view. 3, USNM 7850, plaster cast of the paralectotype (original of Meek, 1877, pl. 13, fig. 1a). 4, USNM 501381, USGS Mesozoic locality D8999 (Text-fig. 11, loc. 36). All × 1.

WALASZCZYK and COBBAN, *Cremnoceramus*

widely spaced (Pl. 19, fig. 4; Pl. 22, fig. 2). Rugae may be symmetrical (Pl. 25, figs 5–6), distinctly asymmetrical (Pl. 23, fig. 3), or may become lamellaete (Pl. 24, fig. 4). Rugae also vary in regularity, presence/absence of intercalated ribs, and evenness of interspaces.

Outline variation results mainly from variation in obliquity. Less oblique forms (Pl. 19, fig. 4; Pl. 21, fig. 4; Pl. 22, fig. 2; Pl. 24, fig. 1; Pl. 25, figs 2, 5–7) have posterior auricle widely extended and distinctly separated from a disc. Posterior auricle is smaller, less distinctly separated in more oblique forms (e.g. Pl. 20, fig. 4; Pl. 21, figs 1–2, 5; Pl. 22, figs 1, 4–5; Pl. 25, fig. 8).

State of preservation has a profound effect on shell outline and general appearance. *C. deformis erectus* was an inflated bivalve. Forms that are flat laterally are either juvenile stages or compressed as a result of sediment compaction. The outline, shape of the anterior margin, and shape of the posterior auricle in these forms change considerably. This is well shown by comparing undeformed specimens in sandstones, including the lectotype (Pl. 19, figs 1–3; Pl. 20, figs 3–4; Pl. 21, fig. 5; Pl. 23, figs 1–2; Pl. 24, fig. 1), with material from marl and clay facies (Pl. 19, fig. 4; Pl. 21, fig. 6; Pl. 22, figs 4–5). These crushed specimens give an impression of roundness with the posterior auricle usually poorly separated. Inflated individuals were referred in Europe to *Cremnoceramus lueckendorfensis* (Tröger, 1967), which, consequently, is regarded here as a synonym of *C. deformis erectus*. Separation on the basis of size is rejected, as in both the Western Interior and Europe; large specimens occur in the basal *erectus* I Event (= *rotundatus* Event) (see Walaszczyk and Wood 1999*a*). Walaszczyk (1992) referred all of these forms to *C. brongniarti* (Mantell, 1822). Although very similar to *C. deformis erectus*, the type of Mantell's species also resembles some representatives of *Tethyoceramus*. Taking into account its doubtful stratigraphical location and the high homeomorphy among inoceramids, an unequivocal statement of its synonymy with Meek's species is impossible (Walaszczyk and Wood 1999*a*) unless it can be determined definitely that its type comes from the corresponding stratigraphical bed.

Another name, which unfortunately appeared recently in the literature (Voigt and Hilbrecht 1997; Callom 1998) is *Cremnoceramus tarlovensis* (see synonymy). This is a *nomen nudum* proposed during the Second Inoceramid Workshop, Freiberg, 1996, as a new name for forms referred to *Cremnoceramus rotundatus* (*sensu* Tröger *non* Fiege), when it appeared that *C. rotundatus* (Fiege) is different and distinctly younger (from the *crassus* Zone) than Tröger's species.

Occurrence. Lower Coniacian, the *C. deformis erectus* Zone; known in North America (Gulf Coast, Western Interior), Europe (Spain, France, England, Germany, Poland, Czech Republic, Romania, Bulgaria, Ukraine, Russia), western Asia (Kazakhstan, Turkmenistan, Caucasus), and Pacific Asia (Japan).

Cremnoceramus deformis dobrogensis (Szász, 1985)

Plate 26, figure 3; Plate 27, figures 2–3

1964 *Inoceramus deformis* Meek; Scott and Cobban, pl. 1.
1965 *Inoceramus deformis* Meek; Hattin, fig. 5.2.
1977 *Inoceramus deformis* Meek; Hattin and Cobban, text-fig. 9–9.
1978 *Inoceramus deformis* Meek; Hattin and Siemers, text-fig. 11–9.
1985 *Inoceramus* aff. *rotundatus* Fiege; Szász, pl. 11, fig. 2.
1985 *Inoceramus dobrogensis* Szász, p. 162 (*pars*), pl. 1, fig. 1; ?pl. 7, fig. 2; pl. 11 fig. 3; ?pl. 24, fig. 1.
1985 *Inoceramus schloenbachi* Böhm; Szász, p. 159 (*pars*); pl. 3, fig. 2; pl. 15, fig. 2; pl. 16, fig. 2; pl. 27, fig. 2.

EXPLANATION OF PLATE 20

Figs 1–4. *Cremnoceramus deformis erectus* (Meek, 1877). 1, USNM 501414, USGS Mesozoic locality 20611 (Text-fig. 11, loc. 42): a, lateral view; b, anterior view. 2, USNM 501409, USGS Mesozoic locality 21421 (Text-fig. 11, loc. 16). 3, USNM 501412, USGS Mesozoic locality D8284 (Text-fig. 11, loc. 80): a, lateral view of the RV; b, posterior view. 4, USNM 501411, USGS Mesozoic locality D8284 (Text-fig. 11, loc. 80): a, lateral view of RV; b, posterior view. All ×1.

PLATE 20

WALASZCZYK and COBBAN, *Cremnoceramus*

1985 *Inoceramus pseudoinconstans* Szász, p. 157 (*pars*), pl. 26, fig. 1.
1985 *Inoceramus paradeformis* Szász, p. 165 (*pars*), pl. 16, fig. 1; pl. 17, fig. 1;

Type. The holotype, by original designation (Szász 1985, p. 162, pl. 1, fig. 1), specimen LPB-III 1-0351, in the collections of the University of Bucharest, Romania, from the Lower Coniacian of the Babadag Basin, Dobrogea, Romania.

Material. Numerous specimens in the collections of the Geological Survey, Denver.

Description. Moderate to large size for genus, inequilateral, equivalved. Valves moderately to strongly inflated, prosocline, usually weakly oblique. Juveniles weakly or moderately inflated. Valve outline subrounded to subquadrate, rarely slightly axially elongated. Beak massive, projecting slightly above the hinge line. Posterior auricle usually small, subtriangular, well separated from massive disc. Anterior margin long, convex, passing into widely rounded ventral margin. Posterior margin slightly convex. Hinge line of moderate size, straight.

Juvenile covered with distinct, strong, concentric rugae, subevenly and subregularly spaced, with very gradual increase of interspaces towards the ventral margin. Adult smooth or bears irregular rugae.

Remarks. The subspecies *dobrogensis* is very close to the nominate form, except for ornament. Interspaces between rugae in juveniles increase regularly ventralward, whereas they are more or less equal in *dobrogensis* subspecies. Although no statistically valid data can be obtained from the material available, *dobrogensis* seems to be smaller in average size as well as more isometric (with length and height equal) than the nominate subspecies.

Inoceramus paradeformis Szász (1985, p. 165, pl. 16, fig. 1; pl. 17, fig. 1), placed close to *C. deformis* by Szász, is a synonym. It differs from *dobrogensis* in being only weakly geniculate. Geniculation (as well as its presence/absence) is a very variable character in *Cremnoceramus*, and does not provide a basis for taxonomic differentiation.

As with *C. deformis deformis*, some specimens of *C. deformis dobrogensis* are characterized by a strongly extended adult stage (see Text-figs 18–20). Such forms were sometimes referred to *Inoceramus browni* Cragin, 1889 (see Frey 1972, pl. 10, figs 7–9; Hattin 1975, pl. 3, fig. 8; Kauffman 1977*b*, pl. 12, fig. 5; Kauffman *et al.* 1978, pl. 16, fig. 5); *I. browni* Cragin is a *nomen nudum*, as pointed out previously by Miller (1968).

The large specimen illustrated by Scott and Cobban (1964, pl. 1), and identified as *Inoceramus deformis*, is an example of these large forms of *C. deformis dobrogensis*. Seitz (1965, p. 130) referred this specimen to *Inoceramus ernsti* Heinz, 1928, a *Tethyoceramus* with quite different early ornament and subsequent ontogeny.

Occurrence. Middle Lower Coniacian, *deformis dobrogensis* and lower *crassus crassus* zones) of North America (Western Interior, Gulf Coast) and Europe (Germany, Poland, Romania, Ukraine, Russia). Well represented in the upper part of the Fort Hays Limestone Member and in the overlying shale and limestone unit of the Smoky Hill Member at Pueblo.

EXPLANATION OF PLATE 21

Figs 1–6. *Cremnoceramus deformis erectus* (Meek, 1877). 1, USNM 501393, USGS Mesozoic locality 20611 (Text-fig. 11, loc. 42). 2, USNM 501394, USGS Mesozoic locality 21421 (Text-fig. 11, loc. 16). 3, USNM 501395, USGS Mesozoic locality D7268 (Text-fig. 11, loc. 67). 4, USNM 501398, USGS Mesozoic locality D8999 (Text-fig. 11, loc. 36). 5, USNM 501396, USGS Mesozoic locality D8284 (Text-fig. 11. Loc. 80): a, lateral view of the LV; b, anterior view. 6, USNM 501397, USGS Mesozoic locality D7268 (Text-fig. 11, loc. 67). All ×1.

WALASZCZYK and COBBAN, *Cremnoceramus*

Cremnoceramus deformis deformis (Meek, 1871)

Plate 26, figures 1, 5; Plate 27, figures 4–5

?1845	*Inoceramus*?; Hall, p. 310, pl. 4, fig. 2.
1871	*Inoceramus deformis* Meek, p. 296.
1877	*Inoceramus deformis* Meek; Meek, p. 146, pl. 14, fig. 4.
1877	*Inoceramus deformis* Meek; White, p. 179, pl. 15, fig. 1.
1894	*Inoceramus deformis* Meek; Stanton, p. 85 (*pars*), pl. 15, fig. 2 [*non* pl. 14, fig. 1; pl. 15, fig. 1 = *Cremnoceramus deformis erectus* (Meek)].
1896	*Inoceramus deformis* Meek; Gilbert, pl. 60, figs 1–2.
1898	*Inoceramus deformis* Meek; Logan, p. 486 (*pars*), pl. 96, fig. 2.
1898	*Inoceramus deformis* Meek; Hill and Vaughan, pl. 61, fig. 2.
1900	*Inoceramus deformis* Meek; Herrick and Johnson, pl. 37, fig. 1.
1910	*Inoceramus deformis* Meek; Grabau and Shimer, p. 443, fig. 582a.
1955	*Inoceramus deformis* Meek; Cobban, p. 207, pl. 3.
1964	*Inoceramus deformis* Meek; Scott and Cobban, pl. 1.
non 1967	*Inoceramus deformis* Meek; Tröger, p. 130, pl. 14, fig. 7 [= *Cremnoceramus crassus crassus* (Petrascheck, 1903)].
1968	*Inoceramus deformis* Meek; Miller, p. 23, pl. 4, figs 1–3.
1970	*Inoceramus deformis* Meek; Miller, p. 237, pl. 2, fig. 2.
1972	*Inoceramus deformis* Meek; Frey, p. 29, pl. 9, figs 4, 7; pl. 13, fig. 9.
1974	*Inoceramus deformis* Meek; Tröger, p. 114.
1977*b*	*Inoceramus* (?) *deformis* Meek, n. ssp. ('late form'); Kauffman, pl. 12, fig. 2.
1977*b*	*Inoceramus* (?) *deformis deformis* Meek; Kauffman, pl. 12, fig. 3.
1978	*Inoceramus* (?) *deformis* Meek, n. ssp. ('late form'); Kauffman *et al.*, pl. 16, fig. 2.
1978	*Inoceramus* (?) *deformis deformis* Meek; Kauffman *et al.*, pl. 16, fig. 3.
1979	*Inoceramus deformis* Meek; Ivannikov, p. 48, pl. 6, figs 2–3.
1985	*Inoceramus deformis* Meek; Szász, p. 161, pl. 6, figs 1–2.
1985	*Inoceramus schloenbachi* Böhm; Szász, p. 159 (*pars*), ?pl. 3, fig. 1; pl. 28, fig. 1.
1985	*Inoceramus paradeformis* Szász, p. 165 (*pars*), pl. 4, fig. 1.
1985	*Inoceramus dobrogensis* Szász, p. 162 (*pars*), ?pl. 26, fig. 2; ?pl. 31, fig. 1.
non 1991	*Cremnoceramus deformis* (Meek); Collom, pl. 7, fig. 5; pl. 8, fig. 3.
1992	*Cremnoceramus deformis* (Meek); Walaszczyk, p. 52, pl. 29, fig. 4; pl. 30, fig. 4.
1998	*Cremnoceramus deformis* (Meek); Collom (*pars*), figs 4E, 6A–B, E–F (*non* 6C–D), 8.

Type. The lectotype, designated by Collom (1998, p. 126), is USNM 4460a, illustrated by Meek (1877, pl. 14, fig. 4; re-illustrated here, Pl. 26, fig. 5) from what is now the Fort Hays Limestone Member of the Niobrara Formation in south-east Colorado. Meek based his *Inoceramus deformis* on several specimens, of which only the lectotype has been recognized. The specimen illustrated by Hall (1845, pl. 4, fig. 2), and referred to by Meek in his synonymy, is a syntype, now paralectotype of the species, following Collom's lectotype designation.

The date of description of *C. deformis* is 1871, and not 1877 as accepted by Collom (1998). Although Meek only listed his new species in his paper of 1871, it is accompanied by a valid indication in the form of a reference to Hall's figure (1845, pl. 4, fig. 2).

Material. Numerous specimens in the USGS collections from many localities in the Western Interior.

EXPLANATION OF PLATE 22

Figs 1–6. *Cremnoceramus deformis erectus* (Meek, 1877). 1, USNM 501384, USGS Mesozoic locality D7268 (Text-fig. 11, loc. 67). 2, USNM 501385, USGS Mesozoic locality 3312 (Text-fig. 11, loc. 41). 3, USNM 501383, USGS Mesozoic locality 21421 (Text-fig. 11, loc. 16). 4, USNM 501386, USGS Mesozoic locality D7268 (Text-fig. 11, loc. 67). 5, USNM 501382, USGS Mesozoic locality D7268 (Text-fig. 11, loc. 67). 6, USNM 501388, USGS Mesozoic locality 21421 (Text-fig. 11, loc. 16). All ×1.

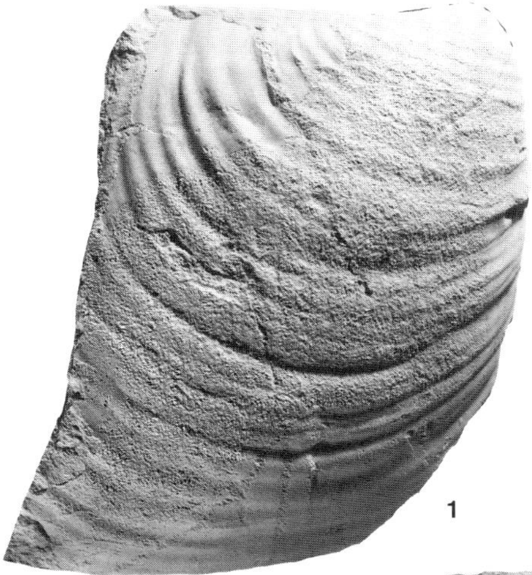

1

2

3

4

5

6

WALASZCZYK and COBBAN, *Cremnoceramus*

Diagnosis. Medium to large-sized *Cremnoceramus*, inequilateral, equivalved, with subquadrate to subrounded outline, strongly inflated. Juvenile covered with widely spaced, sharp-edged rugae, with interspaces increasing in width ventralward. Adult smooth or with very irregular rugae.

Description. Medium to large size, inequilateral, equivalved, highly inflated. Valves prosocline with variable obliquity, usually 70–80 degrees. Beak projecting above the hinge line. Posterior auricle variably developed. Less oblique forms usually much larger, subtriangular in shape, with well-developed auricular sulcus. More oblique specimens usually smaller, elongated. Anterior margin long, convex, passing into widely convex ventral margin, and thence into more or less straight posterior margin. Anterior face steep, high. Juvenile moderately inflated.

Juvenile ornament of distinct concentric rugae with growth marks at edges, subregular, with wide interspaces, increasing gradually (but not evenly) ventrally. Interspaces with irregular ribs. Adults ornamented by irregular low rugae, with very wide interspaces, smooth, with growth lines only.

Remarks. *C. deformis deformis* includes subquadrate forms with rugae, and with interspaces widening gradually ventrally, but not evenly. Forms with subeven spacing of juvenile rugae are here referred to *C. deformis dobrogensis* (Szász, 1985) (see Text-fig. 22).

Occurrence. *C. crassus* Zone, or *crassus/deformis* Zone of North America (Gulf Coast, Western Interior) and Europe (England, France, Germany, Czech Republic, Romania, Poland, Ukraine, Russia). Our material indicates that the nominate subspecies occurs later than that of *C. deformis dobrogensis*, at or above the base of the *crassus* Zone, whereas the subspecies *dobrogensis* occurs below, in an interval formerly referred to as the *deformis* Zone. In Europe *C. deformis deformis* appears in the *deformis* Event, the level that was formerly accepted as the base of the *deformis* Zone (see Ernst *et al*. 1983), which now falls in a lower part of the *crassus/deformis* Zone (Walaszczyk and Wood 1999*a*).

Cremnoceramus crassus (Petrascheck, 1903) lineage

The *Cremnoceramus crassus* lineage originated from the *waltersdorfensis* lineage at the level of the *inconstans* Event (Walaszczyk and Wood 1999*a*; Text-fig. 21), as revealed by the record in Europe. It marks the base of the European *inconstans* Zone, corresponding to the middle part of the *deformis* Zone as defined here and applicable in most of the US Western Interior. The *crassus* lineage comprises two chronosubspecies; *crassus inconstans*, in the upper part of the *deformis* Zone, and *crassus crassus*, above.

Cremnoceramus crassus crassus (Petrascheck, 1903)

Plate 26, fig. 4; Plate 27, figures 6–7; Plate 28, figures 2–6; Plate 29, figures 1, 3; Plate 30, figures 1, 3; Plate 31, fig. 1; Text-figure 25

1843–40 *Inoceramus Cuvieri* Sowerby; Goldfuss, p. 114, pl. 111, fig. 1?a, b, c.
1903 *Inoceramus crassus* Petrascheck, p. 164, pl. 8, fig. 4.
1911 *Inoceramus crassus* Petrascheck; Andert, p. 46, pl. 3, fig. 4; pl. 6, figs 1–2.
1911 *Inoceramus Weisei* Andert, p. 47, pl. 4, figs 2-3; pl. 6, fig. 3.
1912 *Inoceramus schloenbachi* Böhm, p. 570.
1930 *Inoceramus schloenbachi* var. *rostrata* Heinz, p. 28.
1967 *Inoceramus deformis* Meek; Tröger, p. 130, pl. 14, fig. 7.

EXPLANATION OF PLATE 23

Figs 1–4. *Cremnoceramus deformis erectus* (Meek, 1877). 1, USNM 501410, USGS Mesozoic locality D8284 (Text-fig. 13, loc. 80): a, lateral view of RV; b, anterior view. 2, USNM 501391, USGS Mesozoic locality D11939 (Text-fig. 11, loc. 43). 3, USNM 501408, USGS Mesozoic locality D11939 (Text-fig. 11, loc. 43). 4, USNM 501407, USGS Mesozoic locality 21421 (Text-fig. 11, loc. 16). All ×1.

PLATE 23

WALASZCZYK and COBBAN, *Cremnoceramus*

1974 *Inoceramus crassus* Petrascheck; Kotsubinsky, p. 80, pl. 15, fig. 3.
1974 *Inoceramus schloenbachi* Böhm; Tröger, p. 114, pls 1–3.
1975 *Inoceramus inconstans* Woods; Hattin, pl. 3, figs 1, 5.
1985 *Inoceramus crassus anderti* Szász, p. 158.
1986 *Inoceramus (Cremnoceramus) schloenbachi* Böhm; Scott *et al.*, fig. 6m–n.
1991 *Cremnoceramus deformis* (Meek); Collom (*pars*), pl. 8, fig. 3.
1992 *Cremnoceramus crassus* (Petrascheck); Walaszczyk, p. 54, text-fig. 17, pl. 34, figs 1–4; pl. 35, figs 1–22; pl. 36, figs 3, 5.
1992 *Cremnoceramus crassus* (Petrascheck); Čech and Švábenická, pl. 1, fig. 12.
1992 *Cremnoceramus* cf. *schloenbachi*; Čech and Švábenická, pl. 1, fig. 13.
1994 *Cremnoceramus crassus* (Petrascheck); Tröger and Summesberger, p. 168, pl. 1, figs 4–5.
1996 *Cremnoceramus crassus* (Petrascheck); Walaszczyk, p. 380, figs 6C, 7A–B.

Lectotype. By monotypy, specimen figured by Petrascheck (1903, pl. 8, fig. 4; re-illustrated by Walaszczyk 1992, pl. 34, fig. 1) from the Lower Coniacian of Dachsloch quarry near the village of Lešne (German, Innozenzidorf), Bohemia, Czech Republic (see Walaszczyk 1996 for locality map). It is housed in the National Museum, Prague.

Material. Thirty-seven specimens from many localities in the Western Interior.

Description. Inequilateral, equivalved species of moderate to large size. Outline trapezoidal to ovate, elongated parallel to the growth axis. Anterior margin straight or slightly convex, short (30–40 per cent of the relative axial length) passing into broadly rounded, long, anteroventral margin. Posteroventral margin rounded, much shorter. Posterior margin distinctly sulcate. Posterior auricle well developed, well separated from the disc. Juvenile growth axis straight, markedly oblique; angle between growth axis and posterior margin of disc 35–40 degrees on average. Adult part well developed (although rarely preserved); junction with juvenile distinctly geniculate or not. Hinge line long, straight. Ligament plate varies from very high to moderately high, with evenly and uniformly spaced resilifers.

Juvenile valves ornamented with concentric rugae that become more widely spaced ventrally. Rugae covered with raised, lamellaete growth lines up to 30 mm of axial length.

Remarks. *C. crassus crassus* (Petrascheck, 1903) encompasses forms referred to *Inoceramus schloenbachi* Böhm in previous literature (see Walaszczyk 1992). From the morphologically very similar *C. deformis deformis* (Meek), *C. crassus crassus* differs in axial elongation and distinctly greater obliquity, being 55–60 degrees in representatives of the *deformis* lineage (see discussion in Tröger 1974).

C. crassus crassus (Petrascheck) is the younger member of the *crassus* lineage succeeding stratigraphically the subspecies *inconstans*. The subspecies is well represented in the inoceramid record in the Western Interior, as in Europe, and is a very useful marker for the upper part of the Lower Coniacian. Less oblique forms are very close morphologically to *C. deformis deformis*.

Occurrence. *C. crassus crassus* occurs commonly in the upper Lower Coniacian of Europe (England, France, Germany, Spain, Poland, Czech Republic, Romania, Ukraine, Russia), western Central Asia (Kazakhstan, Turkmenistan, Afghanistan), and North America (Gulf Coast, Western Interior). Based on the complete succession in the Staffhorst coal-mine shaft-section in northern Germany (see Niebuhr *et al.* 1999), it is likely that the species does not range into the Middle Coniacian, as defined by the FAD of *Volviceramus koeneni* (Müller, 1887).

EXPLANATION OF PLATE 24

Figs 1–4. *Cremnoceramus deformis erectus* (Meek, 1877). 1, USNM 501389, USGS Mesozoic locality D8284 (Text-fig. 11, loc. 80): a, lateral view of LV; b, posterior view. 2, USNM 501392, USGS Mesozoic locality D6625 (Text-fig. 11, loc. 51). 3, USNM 501390, USGS Mesozoic locality 20611 (Text-fig. 11, loc. 42): a, lateral view of the LV; b, anterior view. 4, USNM 501387, USGS Mesozoic locality 21421 (Text-fig. 11, loc. 16). All ×1.

PLATE 24

WALASZCZYK and COBBAN, *Cremnoceramus*

Cremnoceramus crassus inconstans (Woods, 1912)

Plate 25, figures 3–4; Plate 31, figure 2

1822 *Inoceramus* sp., Mantell, p. 217, pl. 27, fig. 9.
1912 *Inoceramus inconstans* Woods, p. 285 (*pars*), text-figs 42–43, pl. 51, fig. 2.
1977*b* *Cremnoceramus inconstans* (Woods); Kauffman, pl. 13, figs 6, 9–10.
1978 *Cremnoceramus inconstans* (Woods); Kauffman *et al.*, pl. 17, figs 6, 9–10.
1990 *Cremnoceramus inconstans inconstans* (Woods); Kopaevich and Walaszczyk, pl. 4, figs 1–2.
1992 *Cremnoceramus inconstans* (Woods); Walaszczyk, p. 53, pl. 35, fig. 3; pl. 36, fig. 1.
1997 *Mytiloides incertus* (Jimbo); Leckie *et al.* (*pars*), fig. 37Q (*non* fig. 37N–P).

Type. The lectotype, by subsequent designation of Tröger (1967), is the original of Woods (1912, text-fig. 42) from the Upper Chalk of Lewes, England.

Material. Three specimens from the Shelby section and one from the Pueblo section.

Description. Small, markedly oblique, distinctly geniculate, very elongated axially. Anterior margin straight, short, passing into long, broadly rounded anteroventral margin and thence into narrowly rounded ventral margin. Posterior margin short, concave. Hinge line long, straight. Juvenile with subtriangular outline.

Juveniles covered with regularly spaced rugae. Rugae indistinct in umbonal part, where ornament is almost exclusively of raised, sometimes slightly lamellate, growth lines. Adult smooth or with irregular rugae.

Remarks. *C. crassus incostans* is interpreted as the oldest chronosubspecies within the *crassus* lineage, and is usually represented by small individuals. Thus, forms closely resembling Woods' type (see Pl. 27, figs 6–7; Pl. 28, figs 4–6; Pl. 30, fig. 1; Pl. 31, fig. 1; Pl. 32, figs 1–3), but co-occurring with more advanced forms of the lineage, are referred to *C. crassus crassus* (Petrascheck, 1903); these juveniles usually have more widely spaced rugae when compared with early representatives of the lineage (Pl. 25 figs 3–4).

The origin of the *crassus* lineage from the *waltersdorfensis* lineage is well recorded in Europe, where these forms dominate the inoceramid assemblages in the mid-Lower Coniacian (see Walaszczyk and Wood 1999*a*).

Occurrence. *Cremnoceramus crassus inconstans* is known from the middle and upper Lower Coniacian of Europe (England, France, Spain, Germany, Czech Republic, Romania, Poland, Ukraine, Russia) and North America (Gulf Coast, Western Interior). In the Western Interior, *C. crassus inconstans* is very poorly represented; it is known from the upper part of the *deformis* Zone of the Shelby and Pueblo sections.

Genus TETHYOCERAMUS Sornay, 1980

Type species. *Inoceramus* (*Tethyoceramus*) *basseae* Sornay (1980, pl. 1, figs 1, 4, 6; pl. 2, figs 1–3), by original designation.

EXPLANATION OF PLATE 25

Figs 1–2, 5–8. *Cremnoceramus deformis erectus* (Meek, 1877). 1, USNM 501415, USGS Mesozoic locality D11939 (Text-fig. 11, loc. 43). 2, USNM 501416, USGS Mesozoic locality 3312 (Text-fig. 11, loc. 41). 5, USNM 501413, USGS Mesozoic locality 21421 (Text-fig. 11, loc. 16). 6, USNM 501417, USGS Mesozoic locality 21421 (Text-fig. 11, loc. 16). 7, USNM 21119, original of Stanton (1894, pl. 14, fig. 1; pl. 15, fig. 1). 8, USNM 501431, USGS Mesozoic locality D8999 (Text-fig. 11, loc. 36).

Figs 3–4. *Cremnoceramus crassus inconstans* (Woods, 1912). 3, USNM 501419; 4, USNM 501421: USGS Mesozoic locality 21421 (Text-fig. 11, loc. 16).

All ×1.

WALASZCZYK and COBBAN, *Cremnoceramus*

TEXT-FIG. 25. *Cremnoceramus crassus crassus* (Petrascheck, 1903), USNM 501457, USGS Mesozoic locality D7522 (Text-fig. 11, loc. 53); ×1.

Diagnosis. See Walaszczyk and Wood (1999*a*).

Remarks. The group is more poorly represented in the Western Interior than in Europe. In the Old World the group appears at the level of the *erectus* III event (and perhaps slightly lower), and is relatively common in the *hannovrensis* and *inconstans* zones (and probably also in the *crassus* Zone, but the record is

EXPLANATION OF PLATE 26

Figs 1, 5. *Cremnoceramus deformis deformis* (Meek, 1871). 1, USNM 501608, USGS Mesozoic locality D8988 (Text-fig. 11, loc. 59); ×0·55. 5, plaster cast of the holotype, USNM 4460; ×0·66.

Fig. 2. *Cremnoceramus waltersdorfensis hannovrensis* (Heinz, 1932), USNM 501452, USGS Mesozoic locality D3025 (Text-fig. 11, loc. 65); ×1.

Fig. 3. *Cremnoceramus deformis dobrogensis* (Szász, 1985), USNM 501607, USGS Mesozoic locality D8988; ×0·66.

Fig. 4. *Cremnoceramus crassus crassus* (Petrascheck, 1903), USNM 501430, USGS Mesozoic locality 21372 (Text-fig. 11, loc. 19); ×1.

1

2

3

4

5

WALASZCZYK and COBBAN, *Cremnoceramus*

not sufficiently known). In the corresponding interval in the Western Interior (*deformis* Zone), the inoceramid record is dominated by early *C. deformis dobrogensis*. Rare tethyoceramids are represented by *T*? *wandereri* (Andert, 1911), *T. ernsti* (Heinz, 1928) and a form referred here to '*Tethyoceramus alpinus* (Heinz. 1932)'. Moreover, there are some forms referred to as *Tethyoceramus* sp. (Pl. 27, fig. 1; Pl. 28, fig. 1; Pl. 29, fig. 2)

Occurrence. Middle and upper part of the Lower Coniacian.

Tethyoceramus wandereri (Andert, 1911)

Plate 30, figure 2

1911	*Inoceramus Wandereri* Andert, p. 60, pl. 5, fig. 1; pl. 8, fig. 1.
1911	*Inoceramus* cf. *Koeneni* Müller; Andert, p. 60, pl. 5, fig. 3; pl. 8, fig. 2.
1932*b*	*Pleiacoceramus uncinatus* Heinz, p.14.
1934	*Inoceramus wandereri* Andert; Andert, p. 130, text-fig. 15.
1955	*Inoceramus wandereri* Andert; Pavlova, p. 187 (*pars*), pl. 8, figs 1–2; pl. 9, figs 1–2 (*non* pl. 9, figs 3–5).
non 1959	*Inoceramus wandereri* Andert; Dobrov and Pavlova, p. 144, pl. 6, fig. 2; pl. 7, figs 1–2.
1974	*Inoceramus wandereri* Andert; Kotsubinsky, p. 79, pl. 17, figs 1–2.
non 1977*b*	*Cremnoceramus wandereri* (Andert); Kauffman, pl. 13, fig. 11.
non 1978	*Cremnoceramus wandereri* (Andert); Kauffman *et al.*, pl. 17, fig. 11.
?1979	*Inoceramus wandereri* Andert; Ivannikov, p. 87, pl. 30, fig. 3.
1992	*Inoceramus wandereri* Andert; Walaszczyk, pl. 36, fig. 2.
1996	*Inoceramus wandereri* Andert; Walaszczyk, p. 389.

Type. The lectotype, by subsequent designation of Heinz (1932*b*, p. 14), is the unregistered specimen housed in the Museum des Humboldvereins in Ebersbach, Saxony, Germany, illustrated by Andert (1911, pl. 5, fig. 1a; pl. 8, fig. 1a), from Dachsloch, near the village of Lešne (German, Innozenzidorf), Czech Republic (see Walaszczyk 1996 for locality map).

Material. One medium-sized specimen from USGS locality 21273 (USNM 501435; Pl. 30, fig. 2) and 15 small specimens from Shelby, USGS localities D4486 and D12199.

Description. Small to moderate-sized for genus. Valves much higher than long, with slender, pointed umbo projecting very high above the hinge line. Valve outline subtriangular to ovate, elongated parallel to growth axis. Anterior margin long, concave, with anterior wall steep, high. Ventral margin narrowly rounded. Posterior margin convex or straight. Hinge line relatively short, straight. Posterior auricle small, subtriangular in shape, well separated from the disc.

Ornament composed of usually widely, subregularly to irregularly spaced, concentric rugae. Umbonal part often smooth, with growth lines only visible. Rugae best developed in the middle and ventral part of disc.

EXPLANATION OF PLATE 27

Fig. 1. *Tethyoceramus* sp., USNM 501424, USGS Mesozoic locality 21421 (Text-fig. 11, loc. 16), anterior view of RV; ×1.

Figs 2–3. *Cremnoceramus deformis dobrogensis* (Szász, 1985). 2, USNM 501453, USGS Mesozoic locality D13030 (Text-fig. 11, loc. 66); ×0·85. 3, USNM 501454, USGS Mesozoic locality D8988 (Text-fig. 11, loc. 59); ×0·66.

Figs 4–5. *Cremnoceramus deformis deformis* (Meek, 1871). 4, USNM 501456, locality unknown; ×0·6. 5, USNM 501455, USGS Mesozoic locality D12992 (Text-fig. 11, loc. 62); ×0·6

Figs 6–7. *Cremnoceramus crassus crassus* (Petrascheck, 1903). 6, USNM 501433, USGS Mesozoic locality 21422 (Text-fig. 11, loc. 17); 7, USNM 501426, USGS Mesozoic locality 21422 (Text-fig. 11, loc. 17); ×1.

WALASZCZYK and COBBAN, *Cremnoceramus, Tethyoceramus*

Remarks. We refer specimens from localities D12199 and D 4486 to *Tethyoceramus wandereri*, in spite of the fact that these are all internal moulds, and it is impossible to determine whether they had *Cremnoceramus*-type ornament. Forms very close to the small specimens from Shelby were described and illustrated from the Caucasus by Khalafova (1969), and referred by her to a number of new subspecies within *Inoceramus wandereri* Andert (see Khalafova 1969, pl. 16, figs 1–5).

T. wandereri (Andert, 1911) has been used very inconsistently, as indicated by Szász (1985). Many forms referred to this species are actually cremnoceramids with small juvenile and very extended adult parts (as *I. wandereri* in Pavlova, 1955, pl. 9, figs 3–5, and Dobrov and Pavlova 1959, pl. 7, figs 1–2), giving an impression of a slender morphotype similar to the type. This is characterized by a very elongated umbonal part, with a dorsally curved pointed beak, resembling inflated valves of *I. inaequivalvis* rather than *Cremnoceramus* species.

T. wandereri reported from the Western Interior by Kauffman (1977*b*, pl. 13, fig. 11; also *in* Kauffman *et al.* 1978, pl. 17, fig. 11) represents some other species with a distinctive umbonal region that is clearly geniculate, followed by a high adult stage.

Occurrence. T.? *wandereri* (Andert) appears in the *erectus* Zone in the Western Interior and ranges high into the *crassus* Zone. In western and central Europe (Germany, Czech Republic, Poland) *T. wandereri* is very poorly represented, being known by single specimens from the equivalents of the *dobrogensis* and *crassus* zones. The species is well represented in eastern Europe (Ukraine, Russia), where it is commonly reported in association with *C. deformis* and *C. crassus*.

Tethyoceramus ernsti (Heinz, 1928)

Text-figure 26

1896 *Inoceramus brongniarti* Mantell; Inostranzeff, pl. 7, fig. 13.
1928*a* *Inoceramus ernsti* Heinz, p. 73.
1967 *Inoceramus ernsti* Heinz; Tröger, p. 128, pl. 14, figs 1–4, 6.
1979 *Inoceramus ernsti* Heinz; Ivannikov, p. 51, pl. 8, figs 1–2.
non 1980 *Inoceramus* (*Inoceramus*) *ernsti* Heinz; Kauffman, *in* Klinger *et al.*, p. 310, fig. 10G–P.
1985 *Inoceramus ernsti* Heinz; Szász, p. 172, pl. 29, fig. 3.
1991 *Inoceramus ernsti* Heinz; Tarkowski, p. 108, pl. 14, fig. 5.
1992 *Cremnoceramus ernsti* Heinz; Walaszczyk, p. 55 (*pars*), text-fig. 18, pl. 32, fig. 3 (*non* pl. 32, figs 1–2).
1999*a* *Tethyoceramus ernsti* (Heinz); Walaszczyk and Wood, pl. 19, fig. 5.

Type. Lectotype, by subsequent designation of Walaszczyk and Wood (1999*a*), is the original of *Inoceramus brongniarti* Mantell of Inostranzeff (1896, pl. 7, fig. 13) from the Coniacian of the Caucasus. It is housed in the collections of the University Museum of the State University, St Petersburg, Russia.

Material. USNM 501461 from locality 43 (D11939), Frontier Formation, Lower Coniacian *Scaphites preventricosus* ammonite Zone.

Description. Single, large (h max, *c*. 90 mm; l max, 94 mm), highly inflated (B max, 45 mm) double-valved specimen

EXPLANATION OF PLATE 28

Fig. 1. *Tethyoceramus* sp., USNM 501436, USGS Mesozoic locality 21421 (Text-fig. 11, loc. 16); ×1.

Figs 2–6. *Cremnoceramus crassus crassus* (Petrascheck, 1903). 2, USNM 501458, USGS Mesozoic locality 23078 (Text-fig. 11, loc. 63); ×1. 3, USNM 501432, float from one of the upper limestone layers in the shale and limestone unit of the Niobrara Formation, south-west Pueblo; coll. G. R. Scott. 4, USNM 501434, USGS Mesozoic locality 21372 (Text-fig. 11, loc. 19). 5, USNM 501428, USGS Mesozoic locality D9771 (Text-fig. 11, loc. 37). 6, USNM 501427, USGS Mesozoic locality 23280 (Text-fig. 11, loc. 18). 4–6 ×1; 2–3, ×1.

WALASZCZYK and COBBAN, *Cremnoceramus, Tethyoceramus*

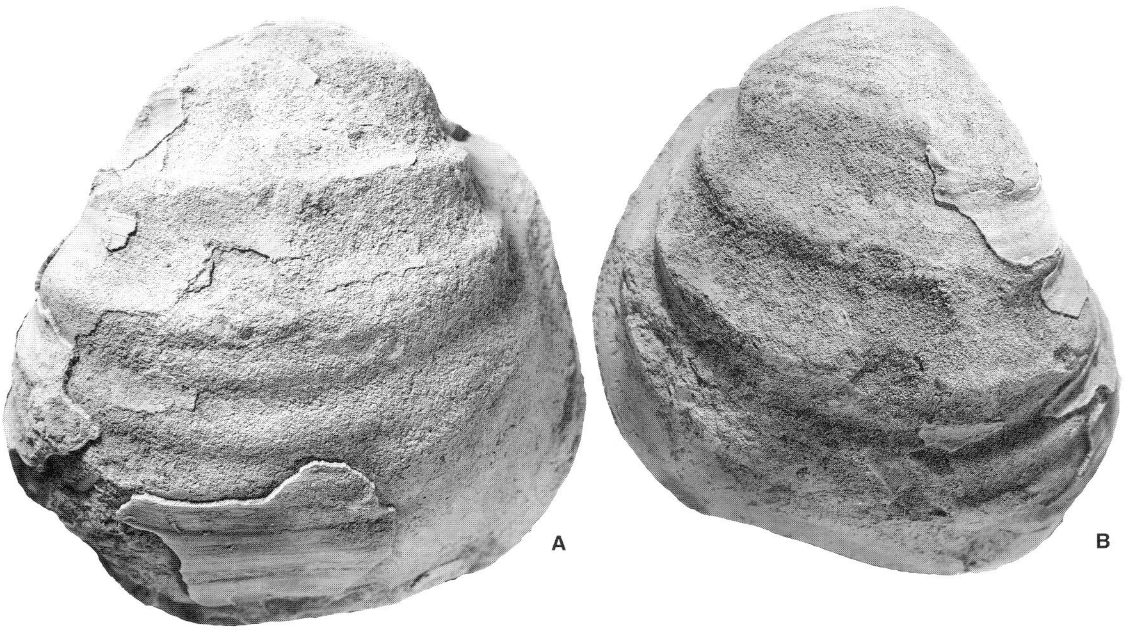

TEXT-FIG. 26. *Tethyoceramus ernsti* (Heinz, 1928). A–B, USNM 501461, USGS Mesozoic locality D11939 (Text-fig. 11, loc. 43); ×0·9.

with parts of prismatic shell layer in the anterior and dorsal parts. Equivalved, with strongly inflated, inequilateral valves. Valves are subtriangular or trapezoidal in outline. Two growth stages can be distinguished. Juvenile (h max, 46·5 mm; l max, 45 mm), moderately inflated, with concave anterior, wide, rounded ventral and almost straight posterior margins. Anterior hinge angle (α) attaining 123 degrees; δ, about 90 degrees. Juvenile covered with regularly and closely spaced, round-topped rugae. Adult part with trapezoidal outline is highly inflated and contacts the juvenile part at nearly a right angle. It is covered with widely spaced rugae with flat interspaces, and growth marks at edges of rugae. Axial part of disc of adult bears weakly developed sulcus. Posterior auricle well separated from disc. Hinge line straight, moderately long.

Remarks. In general shape, growth stages, character of anterior margin, and presence of sulcus in the axial part of the disc in the adult, these specimens correspond closely to the type of *Inoceramus callosus* Heinz, 1932, the original of Woods' (1912) text-figure 85. It differs, however, in possessing *Cremnoceramus*-like ornament, and is accordingly referred to *Tethyoceramus* Sornay (see discussion in Walaszczyk and Wood 1999*a*).

Occurrence. The single specimen represented in the material studied comes from the middle Lower Coniacian locality D11939 (?*dobrogensis* Zone). The species is also known from the middle and upper Lower Coniacian of Europe.

EXPLANATION OF PLATE 29

Figs 1, 3. *Cremnoceramus crassus crassus* (Petrascheck, 1903). 1, USNM 501423; 3, USNM 501425: both USGS Mesozoic locality 21422 (Text-fig. 11, loc. 17).

Fig. 2. *Tethyoceramus* sp., USNM 501424, USGS Mesozoic locality 21421 (Text-fig. 11, loc. 16); lateral view of the RV.

All ×1.

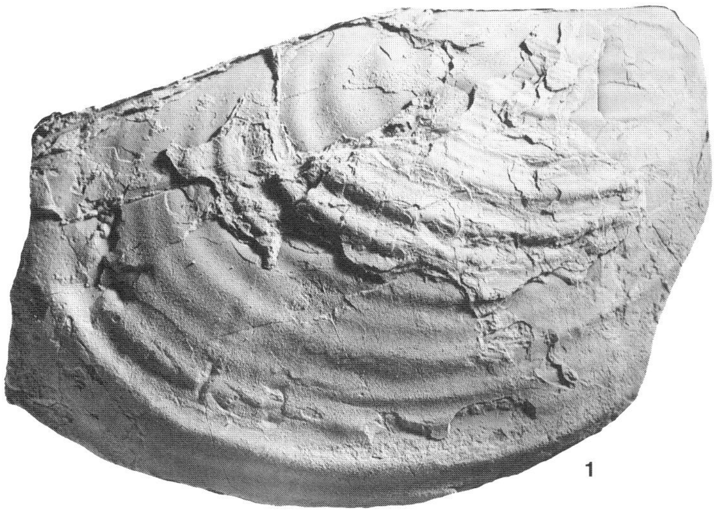

1

2

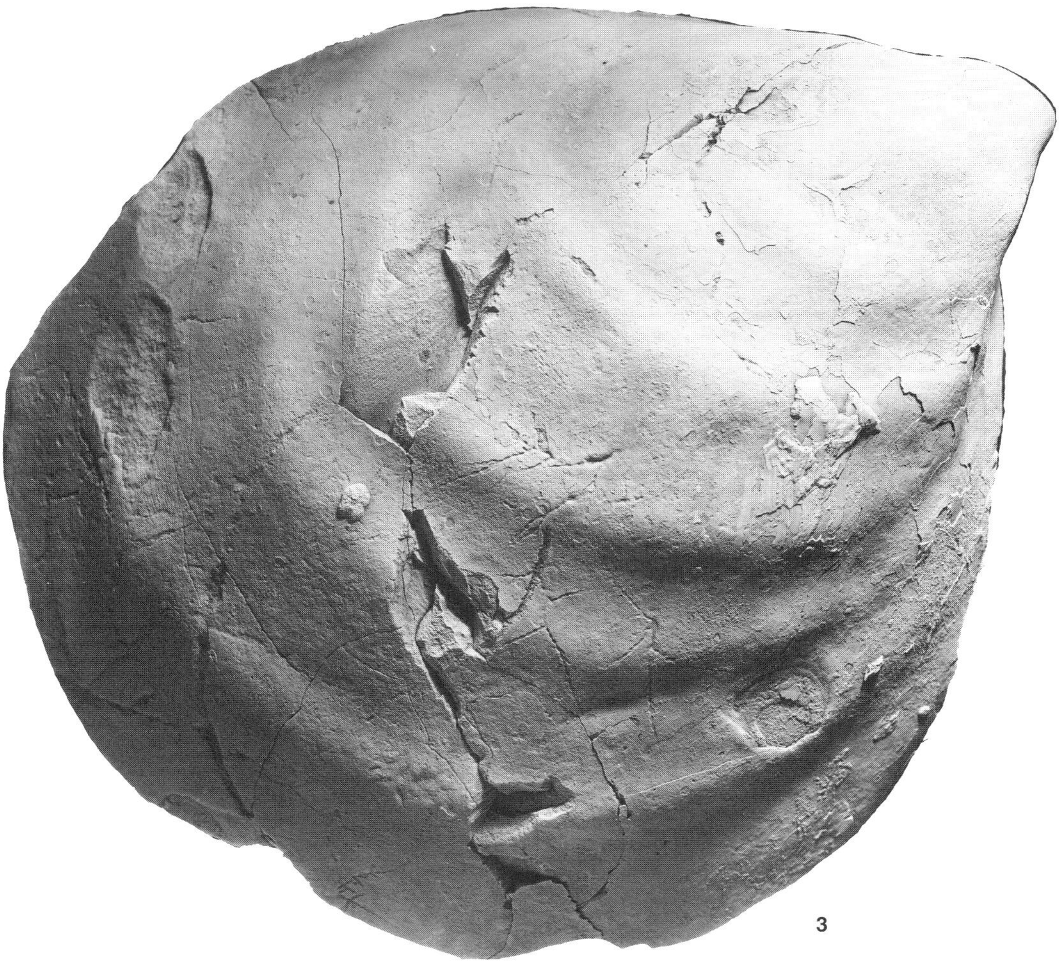

3

WALASZCZYK and COBBAN, *Cremnoceramus, Tethyoceramus*

TEXT-FIG. 27. *Tethyoceramus alpinus* Heinz, 1932, USNM 501462, USGS Mesozoic locality 5865 (Text-fig. 11. Loc. 60); ×0·7.

Tethyoceramus alpinus Heinz, 1932*b*

Text-figure 27

?1904 *Inoceramus cordiformis* Sowerby; Airaghi, p. 189 (*pars*), pl. 4, fig. 9.
?1932*b* *Tethyoceramus* (*Climacoceramus*) *alpinus* Heinz, p. 11.
?1974 *Inoceramus alpinus* Heinz; Kotsubinsky, p. 80, pl. 16, fig. 1

Description. We have a single specimen from the Fort Hays Limestone of Horse Creek, Wyoming. It is a large, undeformed RV. Juvenile high, short, projected markedly above hinge line, almost isometric (with equal height and length); adult pear-shaped in lateral view. Posterior auricle large, triangular, well separated from disc and moderately extended posteriorly in dorsal part. Anterior margin distinctly concave with anterior face steep, high. Ventral margin rounded, passing into straight posterior margin. Hinge line moderately long, straight. Well-developed *Cremnoceramus*-type ornament in the juvenile and middle growth stages, with widely spaced, sharp-edged, concentric rugae with growth marks. Rugae weakly developed on the posterior auricle. Adult with weakly developed rugae, almost smooth.

EXPLANATION OF PLATE 30

Figs 1, 3. *Cremnoceramus crassus crassus* (Petrascheck, 1903). 1, USNM 501429, USGS Mesozoic locality 21422 (text-fig. 11, loc. 17). 3, USNM 501437, USGS Mesozoic locality 23280 (Text-fig. 11, loc. 18).
Fig. 2. *Tethyoceramus wandereri* (Andert, 1911), USNM 501435, USGS Mesozoic locality 21273 (Text-fig. 11, loc. 20): a, lateral view of RV; b, anterior view of the same valve.
All ×1.

WALASZCZYK and COBBAN, *Cremnoceramus, Tethyoceramus*

Remarks. Heinz (1932*b*, p. 11) failed to designate a type for his new species, but referred to one of Airaghi's specimens of *Inoceramus cordiformis* (Airaghi 1904, pl. 4, fig. 9) as being conspecific with his *I. alpinus*, of which it is, therefore, the holotype by monotypy. The Airaghi specimen has not been traced, neither has his other specimen of *I. cordiformis* (Airaghi 1904, pl. 4, fig. 6), the holotype of *Inoceramus novalensis* Heinz (1932*b*, p. 10), which is very close to, or even conspecific with, *T. alpinus*.

Occurrence. As above, probably from the basal *crassus* Zone.

Acknowledgements. We are greatly indebted to Professors J. W. Kennedy (Oxford) and P. J. Harries (Tampa), for their thorough reviews of the manuscript, which have markedly improved the final version of this paper. Similarly, the comments and suggestions of an anonymous referee and the constant help of Dr R. Wood (Cambridge) are gratefully acknowledged. The US Geological Survey kindly made available nearly all of the specimens in this report as well as provided the space and library requirements for the authors. Plaster casts of many types were loaned to us by the Denver Museum of Natural History, Boulder, Colorado via W. D. Bateman. Walaszczyk acknowledges with warm thanks the Fullbright Foundation research fellowship that enabled him to study the North American Upper Cretaceous. We also thank R. E. Burkholder, now retired from the US Geological Survey in Denver, who took most of the photographs.

REFERENCES

AIRAGHI, C. 1904. Inocerami del Veneto. *Bollettino, Società Geologica Italiana*, **23**, 178–199, pl. 1.

ANDERSON, F. M. 1958. Upper Cretaceous of the Pacific Coast. *Memoir of the Geological Society of America*, **71**, xi + 378 pp., 75 pls.

ANDERT, H. 1911. Die Inoceramen des Kreibitz-Zittauer Sandsteingebirges. *Festschrift des Humboldtvereins zur Feier seines 50 jährigen Bestehens*, 33–64, pls 1–9.

—— 1913. *Inoceramus inconstans* Woods und verwandte Arten. *Zentralblatt für Mineralogie und Paläontologie, Jahrgang* **1913**, 295–303.

—— 1934. Die Kreideablagerungen zwischen Elbe und Jeschken Teil III: die Fauna der obersten Kreide in Sachsen, Böhmen und Schlesien. *Abhandlungen der Preußischen Geologischen Landesanstalt, Neue Folge*, **159**, 1–447, pls 1–19.

ANDREWS, D. A. 1944. Geologic and structure contour map of Maverick Springs area, Fremont County, Wyoming. *United States Geological Survey, Oil and Gas Investigations, Preliminary Map* **13**, scale 1: 48,000.

ARZUMANOVA, E. M. 1965. Turonian inoceramid species of Upper Badhyz. *Izvestia Akademii Nauk Turkmenskoy SSR*, **2**, 115–123. [In Russian].

ASSMUS, G. 1963. Stratigraphie und Petrographie des Coniac im östlichen Teil der Halberstädter Mulde. Unpublished Diplomarbeit of the Freiberg Technical University, Germany.

ATABEKIAN, A. A. 1969. On some homonyms in Jurassic and Cretaceous inocermids. *Izvestia Akademii Nauk Armianskoy SSR*, **1**, 3–15. [In Russian].

—— 1974. Inoceramids. 211–424. *In: Atlas of fossil fauna of Armenian SSR*. Izdatelstvo Akademii Nauk Armianskoy SSR, Yerevan. [In Russian].

BASS, N. W. 1926*a*. Geologic investigations in western Kansas, with special reference to oil and gas possibilities – Part 1, the geology of Ellis County, Kansas. *Bulletin of the Geological Survey of Kansas*, **11**, 11–52.

—— 1926*b*. Geologic investigations in western Kansas, with special reference to oil and gas possibilities – Part 2, the geology of Hamilton County, Kansas. *Bulletin of the Geological Survey of Kansas*, **11**, 53–82.

EXPLANATION OF PLATE 31

Fig. 1. *Cremnoceramus crassus crassus* (Petrascheck, 1903), USNM 501420, USGS Mesozoic locality 21372 (Text-fig. 11, loc. 19).

Fig. 2. *Cremnoceramus crassus inconstans* (Woods, 1912), USNM 501418, USGS Mesozoic locality 21421 (Text-fig. 11, loc. 16).

Fig. 3. *Inoceramus annulatus* Goldfuss, 1836, USNM 501422, USGS Mesozoic locality 21421 (Text-fig. 11, loc. 16).

All ×1.

1

2

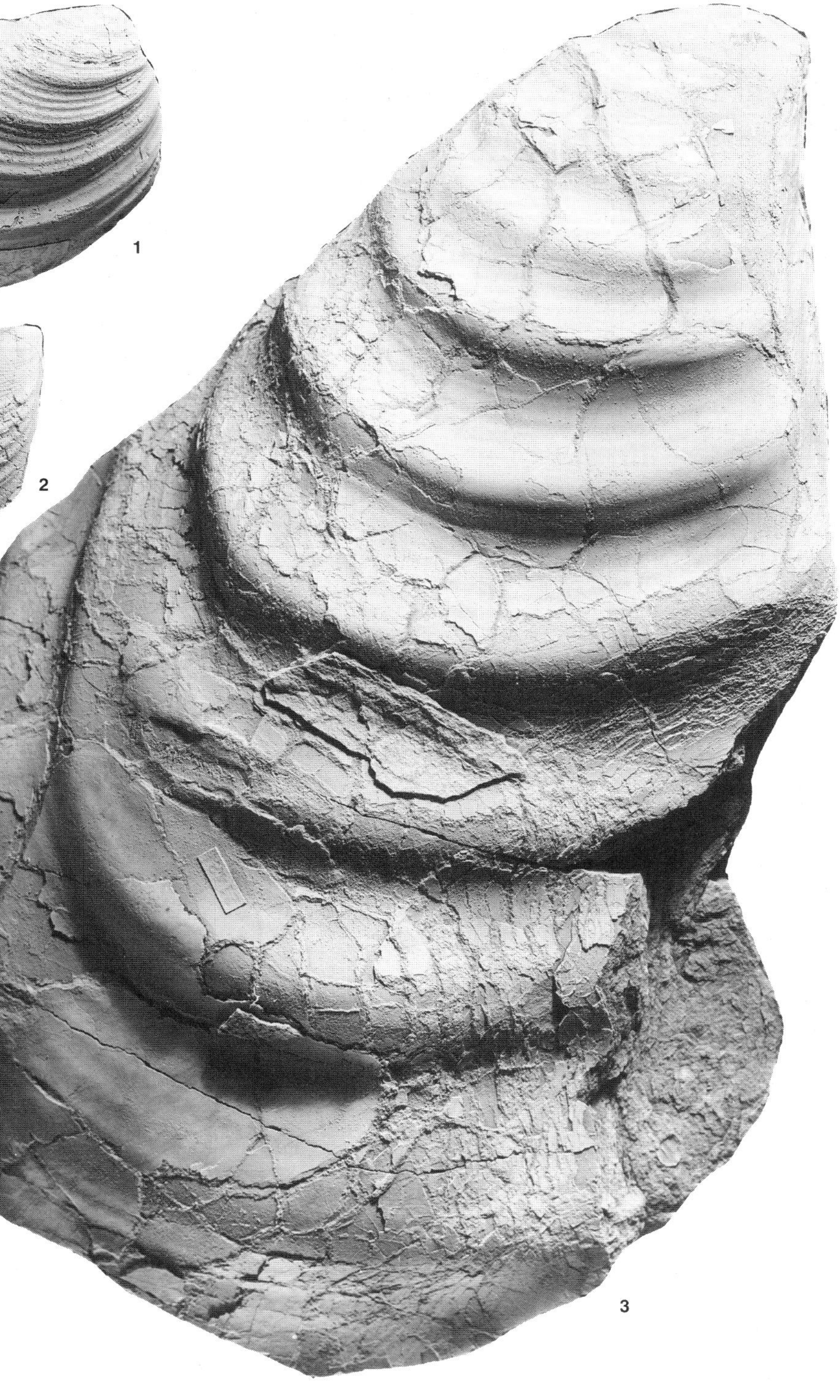
3

WALASZCZYK and COBBAN, *Cremnoceramus, Inoceramus*

BENGTSON, P. (compiler) 1996. The Turonian stage and substage boundaries. *Bulletin de l'Institut Royal des Sciences Naturelles de Begique, Sciences de la Terre*, **66** (Supplement), 69–79.

BODYLEVSKI, W. I. 1958. Upper Cretaceous fauna from the lower reaches of the R. Yenisey. *In* BODYLEVSKI, W. I. and SHULGINA, N. I. Jurassic and Cretaceous faunas from the lower reaches of the Yenisey. *Trudy Vsesoiuznovo Nauchnoissledovatel'skogo Instituta Geologii Arktiki*, **93**, 69–86, pls 43–21. [In Russian].

BÖHM, J. 1912. *Inoceramus Cuvieri* Sowerby. *Zeitschrift der Deutschen Geologischen Gesellschaft*, **63**, 569–570.

BOYLE, C. B. 1893. Catalogue and bibliography of North American Mesozoic Invertebrata. *Bulletin of the United States Geological Survey*, **102**, 1–315.

BRÄUTIGAM, F. 1962. Zur Stratigraphie und Paläontologie des Cenomans und Turons in nordwestlischen Harzvorland. Unpublished PhD thesis, University of Braunschweig, Germany.

ČECH, S. and ŠVÁBENICKÁ, L. 1992. Macrofossil and nannofossil of the type locality of the Březno Formation (Turonian–Coniacian, Bohemia). *Věstnik Českého Geologického Ústavu*, **67**, 311–326, pls 1–4.

COBBAN, W. A. 1951*a*. Scaphitoid cephalopods of the Colorado Group. *Professional Paper of the United States Geological Survey*, **239**, 1–39, 21 pls.

—— 1951*b*. Colorado shale of central and northwestern Montana and equivalent rocks of Black Hills. *Bulletin of the American Association of Petroleum Geologists*, **35**, 2170–2198.

—— 1955. Some guide fossils from the Colorado Shale and Telegraph Creek Formation, northwestern Montana. *Billings Geological Society Guidebook, 6th Annual Field Conference, Sweetgrass Arch-Disturbed Belt, Montana*, 198–207, pls 1–4.

—— 1984*a*. Mid-Cretaceous ammonite zones. Western Interior, United States. *Bulletin of the Geological Society of Denmark*, **33**, 71–89.

—— 1984*b*. Molluscan record from a mid-Cretaceous borehole in Weston County, Wyoming. *Professional Paper of the United States Geological Survey*, **1271**, 1–24, 5 pls.

—— 1986. Upper Cretaceous molluscan record from Lincoln County, New Mexico. 77–89. *In* AHLEN, J. L. and HANSON, M. E. (eds). *Southwest Section of the American Association of Petroleum Geologists, Transactions and Guidebook of 1986 Convention, Ruidoso, New Mexico*, Socorro, 157 pp.

—— 1988. Ammonites in clasts of the Juana Lopez Member of the Carlile Shale (Upper Cretaceous) near Pueblo, Colorado. *Bulletin of the United States Geological Survey*, **1837**, E1–E5.

—— ERDMANN, C. E., LEMKE, R. W. and MAUGHAN, E. K. 1976. Type sections and stratigraphy of the members of the Blackleaf and Marias River formations (Cretaceous) of the Sweetgrass Arch, Montana. *Professional Paper of the United States Geological Survey*, **974**, 1–66.

—— and REESIDE, J. B. JR 1951. Frontier Formation near Sinclair, Carbon County, Wyoming. *Wyoming Geological Association Guidebook, Sixth Annual Field Conference, south-central Wyoming*, 60–65.

—— —— 1952*a*. Frontier Formation, Wyoming and adjacent areas. *Bulletin of the American Association of Petroleum Geologists*, **36**, 1913–1961.

—— —— 1952*b*. Correlation of the Cretaceous formations of the Western Interior of the United States. *Bulletin of the Geological Society of America*, **63**, 1011–1044.

COLLOM, C. J. 1991. High-resolution stratigraphic and paleoenvironmental analysis of the Turonian–Coniacian stage boundary (Late Cretaceous) in the lower Fort Hays Limestone Member, Niobrara Formation, Colorado and New Mexico. Unpublished MSc thesis, Brigham Young University, Utah.

—— 1998. Taxonomy, biostratigraphy, and phylogeny of the Upper Cretaceous bivalve *Cremnoceramus* (Inoceramidae) in the Western Interior of Canada and the United States. 119–142. *In* JOHNSON, P. A. and HAGGART, J. H. (eds). *Bivalves: an eon of evolution – paleobiological studies honoring Norman D. Newell.* University of Calgary Press, Calgary.

CONRAD, T. A. 1858. Observations on a group of Cretaceous fossil shells, found in Tippah County, Miss., with description of fifty-six new species. *Journal of the Academy of Natural Sciences of Philadelphia*, **3**, 323–336.

COX, R. R. 1969. Family Inoceramidae Giebel, 1852. 314–321. *In* MOORE, R. C. (ed.). *Treatise on invertebrate*

EXPLANATION OF PLATE 32

Figs 1–4. *Cremnoceramus crassus crasssus* (Petrascheck, 1903). 1, USNM 501438, USGS Mesozoic locality 23280 (Text-fig. 11, loc. 18). 2, USNM 501439, USGS Mesozoic locality 21372 (Text-fig. 11, loc. 19). 3, USNM 501440, USGS Mesozoic locality D9750 (Text-fig. 11, loc. 38). 4, USNM 501441, USGS Mesozoic locality 21422 (Text-fig. 11, loc. 17). All ×1.

PLATE 32

WALASZCZYK and COBBAN, *Cremnoceramus*

paleontology. Part N. Mollusca 6 (1), Bivalvia. Geological Society of America, Boulder, and University of Kansas Press, Lawrence, 489 pp.

CRAGIN, F. W. 1889. Contributions to the paleontology of the plains, No. 1. *Bulletin of Washburn College Laboratory of Natural History*, **2**, 65–68.

CRAMPTON, J. S. 1988. Comparative taxonomy of the bivalve families Isognomonidae, Inoceramidae, and Retroceramidae. *Palaeontology*, **31**, 965–996.

—— 1996. Inoceramid bivalves from the Late Cretaceous of New Zealand. *Monographs of the Institute of Geological and Nuclear Sciences*, **14**, 1–188.

—— 1999. Ontogenetic variation and inoceramid morphology: a note on early Coniacian *Cremnoceramus bicorrugatus* (Cretaceous Bivalvia). *Acta Geologica Polonica*, **48** (for 1998), 367–376, pls 1–2.

DACQUÉ, E. 1939. Die Fauna der Regensburg-Kalkheimer Oberkreide (mit Ausschluss der Spongien und Bryozoen). *Abhandlungen der Bayerische Akademie der Wissenschaften, Mathematisch-Naturwissenschaftliche Abteilung, Neue Folge*, **45**, 1–281, pls 1–18.

DANE, C. H., COBBAN, W. A. and KAUFFMAN, E. G. 1966. Stratigraphy and regional relationships of a reference section for the Juana Lopez Member, Mancos Shale, in the San Juan Basin, New Mexico. *Bulletin of the United States Geological Survey*, **1224-H**, 1–15.

—— PIERCE, W. G. and REESIDE, J. B. JR 1937. The stratigraphy of the Upper Cretaceous rocks north of the Arkansas River in eastern Colorado. *Professional Paper of the United States Geological Survey*, **186-K**, 207–232.

DE WITT, E., REDDEN, J. A., BUSCHER, D. and WILSON, A. B. 1989. Geologic map of the Black Hills area, South Dakota and Wyoming. *United States Geological Survey, Miscellaneous Investigations Series, Map* **I-1910**, scale 1: 250,000.

DOBROV, S. A. and PAVLOVA, M. M. 1959. Inoceramids. 130–165, pls 1–23. *In* MOSKVIN, M. M. (ed.). *Atlas of the Upper Cretaceous fauna of northern Caucasus and Crimea*. Gostoptechizdat, Moscow, 304 pp. [In Russian].

EGOYAN, V. L. 1955. *The Upper Cretaceous deposits of the southwestern part of the Armenian SSR*. Academy of Sciences of the Armenian SSR, 270 pp. [In Russian].

EICHWALD, E. 1865. *Lethaea Rossica ou Paléontologie de la Russie. Volume 2*, 484–497.

ELDER, W. P. and BOX, S. E. 1992. Late Cretaceous inoceramid bivalves of the Kuskowim Basin, southwestern Alaska, and their implications for basin evolution. *Memoir of the Journal of Paleontology,* **26**, 1–39.

ERNST, G., SCHMID, F. and SEIBERTZ, E. 1983. Event-Stratigraphie im Cenoman und Turon von NW-Deutschland. *Zitteliana*, **10**, 531–554.

FIEGE, K. 1930. Über die Inoceramen des Oberturon mit besonderer Berücksichtigung der im Rheinland und Westfalen Vorkommenden Formen. *Palaeontographica*, **73**, 31–47, pls 4–8.

FREY, R. W. 1972. Paleoecology and depositional environment of Fort Hays Limestone Member, Niobrara Chalk (Upper Cretaceous), west-central Kansas. *Paleontological Contributions of the University of Kansas*, **58**, 1–72.

GARDNER, M. H. and CROSS, T. A. 1994. Middle Cretaceous paleogeography of Utah. 471–502. *In* CAPUTO, M. V., PETERSON, J. A. and FRANCZYK, K. J. (eds). *Mesozoic systems of the Rocky Mountain region, USA*. Rocky Mountain Section, SEPM (Society for Sedimentary Geology).

GEINITZ, H. B. 1871–1875. Das Elbthalgebirge in Sachsen. *Palaeontographica*, **20**, 1–319, pls 1–67 (1–74, pls 1–23, 1871; 95–207, pls 24–45, 1872; 207–236, pls 46–52, 1873; 237–276, pls 53–60, 1874; 277–319, pls 61–67, 1875).

GHAMBASHIDZE, R. A. 1978. Some molluscs from the Upper Cretaceous deposits of Minor Caucasus. *Problemy Geologi Gruzji*, **59**, 172–183, pls 1–3. [In Russian].

GILBERT, G. K. 1896. The underground water of the Arkansas Valley in eastern Colorado. *United States Geological Survey, 17th Annual Report*, **2**, 551–601.

GOLDFUSS, A. 1836 (in 1834–1840). *Petrefakta Germaniae tam ea, quae in museo Universitatis Regiae Borussicae Fridericiae Wilhelmiae Rhenanae servantur quam alia quaecunque in Museis Hoeninghusiano, Muensteriano allisque extant, iconibus et descriptionibus illustrata. Volume 2*. Arnz and Co., Düsseldorf, 69–140, pls 106–118.

GRABAU, A. W. and SHIMER, H. W. 1910. *North American index fossils; invertebrates, volume 2*. A. S. Seiler & Co., New York, 909 pp.

GÜMBEL, C. W. 1868. Verzeichnis in der Sammlung des Geol.-Mineral. Vereins in Regensburg vorfindlischen Versteinerungen aus den Schichten der Procän- oder Kreideformation aus der Umgebung von Regensburg. *Correspondenz-Blatt des Zoologisch-Mineralogischen Vereins in Regensburg*, **22**, 1–69.

HALE, L. A. 1960. Frontier Formation – Coalville, Utah and nearby areas of Wyoming and Colorado. *Wyoming Geological Association Guidebook, 15th Annual Field Conference*, 137–146.

—— 1976. Geology of the Coalville anticline, Summit County, Utah. 381–386. *In*: *Symposium on geology of the Cordilleran hingline*. Rocky Mountain Association of Geologists, Denver.

HALL, J. 1845. Descriptions of organic remains collected by Captain J. C. Frémont, in the geographical survey of Oregon and North California. 304–310. *In*: *Report of the exploring expedition to the Rocky Mountains in the year*

1842 and to Oregon and North California in the years 1843–44 by Brevet Captain J. C. Frémont. Government Printing Office, Washington, DC.

HANCOCK, J. M., KENNEDY, W. J. and COBBAN, W. A. 1993. A correlation of the Upper Albian to basal Coniacian sequences of northwest Europe, Texas and the United States Western Interior. 397–434. *In* CALDWELL W. G. E. and KAUFFMAN, E. G. (eds). Evolution of the Western Interior Basin. *Special Paper of the Geological Association of Canada*, **39**, 680 pp.

HARRIES, P. J., KAUFFMAN, E. G. and CRAMPTON, J. S. (eds) 1996. Lower Turonian Euramerican Inoceramidae: a morphologic, taxonomic, and biostratigraphic overview. A report from the first workshop on Early Turonian inoceramids (Oct. 5–8, 1992) in Hamburg, Germany; organized by Heinz Hilbrecht and Peter Harries. *Mitteilungen aus dem Geologisch-Paläontologischen Institut der Universität Hamburg* **77**, 641–671.

HATTIN, D. E. 1962. Stratigraphy of the Carlile Shale (Upper Cretaceous) in Kansas. *Bulletin of the Kansas Geological Survey*, **156**, 1–155.

—— 1965. Upper Cretaceous stratigraphy, paleontology, and paleoecology of western Kansas, with a section on Pierre Shale, by W. A. Cobban. *Geological Society of America, Field Conference Guidebook, 18th Annual Meeting*, 69 pp.

—— 1975. Stratigraphic study of the Carlile-Niobrara (Upper Cretaceous) unconformity in Kansas and northeastern Nebraska. *Special Paper of the Geological Association of Canada*, **13**, 195–210.

—— and COBBAN, W. A. 1977. Fourth and fifth days. Upper Cretaceous stratigraphy, paleontology and paleoecology of western Kansas. *The Mountain Geologist*, **14**, 175–217.

—— and SIEMERS, C. T. 1978. Upper Cretaceous stratigraphy and depositional environments of western Kansas. *Kansas Geological Survey, The University of Kansas Guidebook Series*, **3**, 1–102.

HAYDEN, F. V. 1876. *Eighth annual report of the United States Geological and Geographical Survey of the Territories, embracing Colorado and parts of adjacent territories, being a report of progress of the exploration for the year 1874*. Government Printing Office, Washington, DC, 515 pp.

HEINE, F. 1929. Die Inoceramen des mittelwestfälishen Emschers und unteren Untersenons. *Abhandlungen der Preussischen Geologischen Landesanstalt, Neue Folge*, **120**, 1–124, 19 pls.

HEINZ, R. 1926. Beitrag zur Kenntnis der Stratigraphie und Tektonik der oberen Kreide Lüneburgs. *Mitteilungen aus dem Mineralogisch-Geologischen Staatsinstitut, Hamburg*, **8**, 3–109.

—— 1928*a*. Das Inoceramen-Profil der Oberen Kreide Lüneburgs. Mit Anführung der neuen Formen und deren Kennzeichnung. Beiträge zur Kenntnis der oberkretazischen Inoceramen I. *Jahresbericht des Niedersächsischen Geologischen Vereins zu Hannover*, **21**, 64–81.

—— 1928*b*. Über die bisher wenig beachtete Skulptur der Inoceramen-Schale und ihre stratigraphische Bedeutung. Beiträge zur Kenntnis der oberkretazischen Inoceramen IV. *Mitteilungen aus dem Mineralogisch-Geologischen Staatsinstitut, Hamburg*, **10**, 5–39, pls 1–3.

—— 1928*c*. Über die Oberkreide-Inoceramen Süd-Amerikas und ihre Beziehungen zu denen Europas und anderer Gebiete. Beiträge zur Kenntnis der oberkretazischen Inoceramen V. *Mitteilungen aus dem Mineralogisch-Geologischen Staatsinstitut, Hamburg*, **10**, 42–97, pls 1–4.

—— 1930. Zur stratigraphischen Stellung der Sonnenbergschichten bei Waltersdorf i. Sa. (westsüdwestlich von Zittau). Beiträge zur Kenntnis der oberkretazischen Inoceramen IX. *Jahresbericht des Niedersächsischen Geologischen Vereins zu Hannover*, **23**, 25–29.

—— 1932*a*. Zur Gliederung der sächsisch-schlesisch-böhmischen Kreide unter Zugrundelegung der norddeutschen Stratigraphie. Beiträge zur Kenntnis der oberkretazischen Inoceramen X. *Jahresbericht des Niedersächsischen Geologischen Vereins zu Hannover*, **24**, 24–52.

—— 1932*b*. Aus der neuen Systematik der Inoceramen. Beiträge zur Kenntnis der oberkretazischen Inoceramen XIV. *Mitteilungen aus dem Mineralogisch-Geologischen Staatsinstitut, Hamburg*, **13**, 1–26.

—— 1933. Inoceramen von Madagascar und ihre Bedeutung für Kreide-Stratigraphie. Beiträge zur Kenntnis der oberkretazischen Inoceramen XII. *Zeitschrift der Deutschen Geologischen Gesellschaft*, **85**, 241–259.

HERM, D., KAUFFMAN, E. G. and WIEDMANN, J. 1979. The age and depositional environment of the ''Gosau''- Group (Coniacian-Santonian), Brandenberg/Tirol, Austria. *Mitteilungen der Bayerischen Staatssammlung für Paläontologie und Historische Geologie*, **19**, 27–92, pls 1–11.

HERRICK, C. L. and JOHNSON, D. W. 1900. The geology of the Albuquerque sheet. *Bulletin of the New Mexico University*, **2**, 1–67, pls 1–22.

HILL, R. T. and VAUGHAN, T. W. 1898. Geology of the Edwards Plateau and Rio Grande Plain adjacent to Austin and San Antonio, Texas, with reference to the occurrence of underground waters. *United States Geological Survey, 18th Annual Report*, **2**, 193–321.

HOOK, S. C. and COBBAN, W. A. 1980. Some guide fossils in Upper Cretaceous Juana Lopez Member of Mancos and Carlile Shales, New Mexico. 38–49. *In* KOTTLOWSKI, F. E. *et al.*, *New Mexico Bureau of Mines and Mineral Resources, Annual Report* 1978–1979.

—— MOLENAAR, C. M. and COBBAN, W. A. 1983. Stratigraphy and revision of nomenclature of upper Cenomanian to Turonian (Upper Cretaceous) rocks of west-central New Mexico. 7–28. *In* Contributions to mid-Cretaceous paleontology and stratigraphy of New Mexico, Part II. *New Mexico Bureau of Mineral Resources Circular*, **185**.

INOSTRANZEFF, A. 1896. *Au travers de la chaine principale du Caucase. Recherches géologiques la long de la ligne projété du chemin de fer de Vladikavkas-Tfilis au travers du Col de L'Arkhotis*. Edition de la Direktion des Chemins de Fer de l'Etat, Sankt Peterburg, 250 pp., 22 pls.

IVANNIKOV, A. V. 1979. *Inoceramids of the Upper Cretaceous in south-western part of the East European Platform*. Akademia Nauk Ukrainskoy SSR, Institut Geologicheskich Nauk, Kiev, 102 pp. [In Russian].

JELETZKY, J. A. 1970. Cretaceous macrofaunas. 649–662. *In*: *Geology and economic minerals of Canada. Economic Geology Report, 1. Fifth edition*. Department of Energy, Mines and Resources, Ottawa, Canada.

JERZYKIEWICZ, T. 1969. Old palaeontological evidence of the stratigraphic position of the youngest Upper Cretaceous sandstone (Gory Stolowe, Middle Sudetes). *Bulletin de l'Academie Polonaise des Sciences, Série des Sciences Géologiques et Géographiques*, **17**, 173–176, pls 1–2.

JIMBO, K. 1894. Beiträge zur Kenntnis des Fauna der Kreideformation von Hokkaido. *Paläontologische Abhandlungen, Neue Folge*, **2**, 140–198, pls 1–9.

JOHNSON, D. W. 1903. The geology of the Cerrillos Hills, New Mexico. Part II, Palaeontology. *School of Mines Quarterly*, **24**, 101–145.

KAPLAN, U. 1986. Ammonite stratigraphy of the Turonian of NW-Germany. *Newsletters on Stratigraphy*, **17**, 9–20.

—— and KENNEDY, W. J. 1994. Ammoniten des westfälischen Coniac. *Geologie und Paläontologie in Westfalen*, **31**, 1–155.

—— —— 1996. Upper Turonian and Coniacian ammonite stratigraphy of Westphalia, NW Germany. *Acta Geologica Polonica*, **46**, 305–352.

KAUFFMAN, E. G. 1975. Dispersal and biostratigraphic potential of Cretaceous benthonic bivalvia in the Western Interior. *Special Paper of the Geological Association of Canada*, **13**, 163–194.

—— 1977*a*. Systematic, biostratigraphic, and biogeographic relationships between middle Cretaceous Euramerican and North Pacific Inoceramidae. *Special Papers of the Palaeontological Society of Japan*, **21** (Mid-Cretaceous Events, Hokkaido Symposium), 169–212.

—— 1977*b*. Illustrated guide to biostratigraphically important Cretaceous macrofossils, Western Interior Basin, USA. *Mountain Geologist*, **14**, 225–274.

—— 1978*a*. An outline of middle Cretacous marine history and inoceramid biostratigraphy in the Bohemian Basin, Czechoslovakia. *Annales du Museum d'Histoire Naturelle de Nice*, **4** (13), 1–12.

—— 1978*b*. South African Middle Cretaceous Inoceramidae. *Annales du Museum d'Histoire Naturelle de Nice*, **4** (17), 1–6.

—— 1978*c*. British Middle Cretacous inoceramid biostratigraphy. *Annales du Museum d'Histoire Naturelle de Nice*, **4** (4), 1–12.

—— 1995. Proposed Turonian–Coniacian boundary stratotype, Wagon Mound, northern New Mexico, USA. *Abstracts, Second International Symposium on Cretaceous Stage Boundaries, Sept. 8–14, 1995*, Brussels, Belgium, 1 p.

—— and BENGTSON, P. 1985. Mid-Cretaceous inoceramids from Sergipe, Brazil; a progress report. *Cretaceous Research*, **6**, 311–315.

—— COBBAN, W. A. and EICHER, D. L. 1978. Albian through Lower Coniacian strata. Biostratigraphy and principal events in Western Interior states. *Annales du Museum d'Histoire Naturelle de Nice*, **4** (23), 1–52.

—— HATTIN, D. E. and POWELL, J. D. 1977. Stratigraphic, paleontologic and paleoenvironmental analysis of the Upper Cretaceous rocks of Cimarron County, northwestern Oklahoma. *Memoir of the Geological Society of America*, **149**, 1–150.

—— KENNEDY, W. J. and WOOD, C. J. 1996. The Coniacian stage and substage boundaries. *Bulletin de l'Institut Royal des Sciences Naturelles de Begique, Sciences de la Terre*, **66** (Supplement), 81–94.

—— and PRATT, L. M. 1985 (coordinators). A field guide to the stratigraphy, geochemistry, and depositional environments of the Kiowa-Skull Creek, Greenhorn, and Niobrara marine cycles in the Pueblo-Canyon City area, Colorado. FRS1–FRS26. *In* PRATT, L. M., KAUFFMAN, E. G. and ZELT, F. B. (eds). Fine-grained deposits and biofacies of the Cretaceous Western Interior seaway: evidence of cyclic sedimentary processes. *Society of Economic Paleontologists and Mineralogists, Field Trip Guidebook No. 4, 1985 Midyear Meeting, Golden, Colorado*. Society of Economic Paleontologists and Mineralogists, Tulsa, 249 + FRS1–FRS26 pp.

—— SAGEMAN, B. B., KIRKLAND, J. I., ELDER, W. P., HARRIES, P. J. and VILLAMIL, T. 1993. Molluscan biostratigraphy of the Cretaceous Western Interior Basin, North America. 397–434. *In* CALDWELL, W. G. E. and KAUFFMAN, E. G. (eds). Evolution of the Western Interior Basin. *Special Paper of the Geological Association of Canada*, **39**, 680 pp.

KELLER, S. 1982. Die Oberkreide der Sack-Mulde bei Alfeld (Cenoman–Unter-Coniac). Lithologie, Biostratigraphie und Inoceramen. *Geologisches Jahrbuch A*, **64**, 3–171.

KENNEDY, W. J. 1984. Systematic palaeontology and stratigraphic distribution of the ammonite faunas of the French Coniacian. *Special Papers in Palaeontology*, **31**, 1–160.

—— and COBBAN, W. A. 1991. Coniacian ammonite faunas from the United States Western Interior. *Special Papers in Palaeontology*, **45**, 1–96.

—— —— HANCOCK, J. M. and HOOK, S. C. 1989. Biostratigraphy of the Chispa Summit Formation at its type locality: a Cenomanian through Turonian reference section for Trans-Pecos Texas. *Bulletin of the Geological Institutions of the University of Uppsala, New Series*, **15**, 39–119.

KHALAFOVA, R. A. 1969. *Fauna and stratigraphy of the Upper Cretaceous deposits of the SE part of the Small Caucasus and Nachitchewan area of ASSR*. ASS Academy of Sciences, Yerevan, 330 pp. [In Russian].

KHOMENTOVSKY, O. V. 1998. Inoceramidae (Bivalvia) and biostratigraphy of the Upper Cretaceous of northern Siberia. *Autoreferat*, 1–20.

KIRKLAND, J. I. 1996. Paleontology of the Greenhorn cyclothem (Cretaceous: late Cenomanian to middle Turonian) at Black Mesa, northeastern Arizona. *Bulletin of the New Mexico Museum of Natural History and Science*, **9**, 1–131, pls 1–50.

KLINGER, H. C., KAUFFMAN, E. G. and KENNEDY, W. J. 1980. Upper Cretaceous ammonites and inoceramids from the off-shore Alphard Group of South Africa. *Annals of the South African Museum*, **82**, 293–320.

KNECHTEL, M. M. and PATTERSON, S. H. 1962. Bentonite deposits of the northern Black Hills district, Wyoming, Montana, and South Dakota. *Bulletin of the United States Geological Survey*, **1082-M**, 893–1030.

KOPAEVICH, L. F. and WALASZCZYK, I. 1990. An integrated inoceramid-foraminiferal biostratigraphy of the Turonian and Coniacian strata in south-western Crimea, Soviet Union. *Acta Geologica Polonica*, **40**, 83–96, pls 1–5.

KOTSUBINSKY, S. P. 1958. *Inoceramids of the Cretaceous deposits of the Volhynian-Podolian Plate*. Akademia Nauk Ukrainskoy SSR, Kiev, 49 pp. [In Ukrainian].

—— 1965. A new species of *Inoceramus* from the Upper Turonian of the Volyn-Podolian Plateau. *Paleontological Journal*, **1965** (2), 48–50. [In Russian].

—— 1968. Inoceramidae. 115–148, pls 16–29. *In*: *Fauna of the Cretaceous deposits of the Western Ukraine*. Naukovaya Dumka, Kiev, 272 pp. [In Ukrainian].

—— 1974. Inocerams. 76–86, pls 13–24. *In* KRYMGOLTZ, G. J. (ed.). *Atlas of the Upper Cretaceous fauna of Donbass*. Nedra, Moscow, 380 pp. [In Russian].

KÜCHLER, T. and ERNST, G. 1989. Integrated biostratigraphy of the Turonian-Coniacian transition interval in northern Spain with comparisons to NW Germany. 161–190. *In* WIEDMANN, J. (ed.). *Cretaceous of the Western Tethys. Proceedings, 3rd International Cretaceous Symposium, Tübingen 1987*. E. Schweizerbart'sche Verlagsbuchhandlung, Stuttgart, 1005 pp.

LANDIS, E. R. and DANE, C. H. 1967. Geologic map of the Tierra Amarilla quadrangle, Rio Arriba County, New Mexico. *New Mexico Bureau of Mines and Mineral Resources, Geologic Map*, **19**, scale 1:62,500.

LARSON, P. A., MORIN, R. W., KAUFFMAN, E. G. and LARSON, A. 1991. *Sequence stratigraphy and cyclicity of Lower Austin/Upper Eagle Ford outcrops (Turonian–Coniacian), Dallas County, Texas*. Field Trip No. 9, Dallas Geological Society, Dallas, Texas, 61 pp.

LECKIE, R. M., KIRKLAND, J. I. and ELDER, W. P. 1997. Stratigraphic framework and correlation of a principal reference section of the Mancos Shale (Upper Cretaceous), Mesa Verde, Colorado. *New Mexico Geological Society Guidebook, 48th Field Conference, Mesozoic geology and paleontology of the Four Corners region*, 163–216.

LOGAN, W. N. 1897. The Upper Cretaceous of Kansas. *University Geological Survey of Kansas*, **2**, 195–234.

—— 1898. The invertebrates of the Benton, Niobrara and Fort Pierre Groups. *University Geological Survey of Kansas*, **4** (Paleontology) (Upper Cretaceous **8**), 431–518, pls 86–120.

LONGMAN, N. W., LUNEAU, B. A. and LANDON, S. M. 1998. Nature and distribution of Niobrara lithologies in the Cretaceous Western Interior seaway of the Rocky Mountain Region. *The Mountain Geologist*, **35**, 137–170.

LUCAS, S. G. and ESTEP, J. W. 1998. Cretaceous stratigraphy and biostratigraphy in the southern San Andreas Mountains, Doña Ana County, New Mexico. *New Mexico Geological Society Guidbook, 49th Field Conference, Las Cruces County*, **11**, 187–196.

LUPU, D. 1976. Inocerami din Turonianul si Coniacianul de la Bretelin (Stratele de Deva). *Academie Romana Studii si Cercetări Geologie, Geofizică, Geographie, Seria Geologia*, **21**, 131–141.

MANTELL, G. 1822. *Fossils of the South Downs; or illustrations of the geology of Sussex*. Lupton Relfe, London, 320 pp., 42 pls.

MATHER, K. F., GILLULY, J. and LUSK, R. G. 1928. Geology and oil and gas prospects of northern Colorado. *Bulletin of the United States Geological Survey*, **796-B**, 65–124.

MATSUMOTO, T. 1981. The specimens of *Inoceramus* (Bivalvia) donated to the Kitakyushu Museum of Natural History. *Bulletin of Kitakyushu Museum of Natural History*, **3**, 15–26.

—— 1984. The so-called Turonian-Coniacian boundary in Japan. *Bulletin of the Geological Society of Denmark*, **33**, 171–181.

—— HAYASAMI, I. and ASANO, K. 1963. A survey of fossils from Japan illustrated in classical monographs. *Palaeontological Society of Japan, 25th Anniversary Volume*, 60–68.

—— and NODA, M. 1983. Restudy of *Inoceramus incertus* Jimbo with special reference to its biostratigraphic implication. *Proceedings of Japanese Academy of Sciences, Series B*, **59**, 109–112.

—— —— 1984. A note on an inoceramid species (Bivalvia) from the Lower Coniacian (Cretaceous) of Hokkaido. *Transactions and Proceedings of the Palaeontological Society of Japan, New Series*, **140**, 263–273, pls 41–44.

McLEARN, F. H. 1926. New species from the Coloradoan of lower Smoky and lower Peace Rivers, Alberta. *Bulletin of the Canadian Department of Mines, Geological Survey, Geological Series*, **45**, 117–126.

MEEK, F. B. 1871. Preliminary paleontological report, consisting of lists of fossils, with description of some new types, etc. *United States Geological Survey of Wyoming (Hayden), Preliminary Report*, **4**, 287–318.

—— 1876. A report on the invertebrate Cretaceous and Tertiary fossils of the upper Missouri County. *United States Geological Survey of the Territories (Hayden) Report*, **9**, lxiv + 629 pp, 45 pls.

—— 1877. Paleontology. *United States Geological Survey, Exploration of the Fortieth Parallel*, **4**, 1–197, pls 1–17.

—— and HAYDEN, F. V. 1861. Systematic catalogue, with synonyma, etc., of Jurassic, Cretaceous and Tertiary fossils collected in Nebraska, by the exploring expeditions under the command of Lieut. G. K. Warren, of U.S. Topographical Engineers. *Proceedings of the Academy of Natural Sciences of Philadelphia*, **12**, 417–432.

MEREWETHER, E. A. and COBBAN, W. A. 1986. Biostratigraphic units and tectonism in the mid Cretaceous foreland of Wyoming, Colorado and adjoining areas. *In* PETERSON, J. A. (ed.). Paleotectonism and sedimentation in the Rocky Mountain region. *Memoir of the American Association of Petroleum Geologists*, **41**, 443–468.

MILLER, H. W. 1968. Invertebrate fauna and environment of deposition of the Niobrara Formation (Cretaceous) of Kansas. *Fort Hays Studies, Science Series*, **8**, 1–90.

—— 1970. Additions to the fauna of the Niobrara Formation of Kansas. *Transactions of the Kansas Academy of Science*, **72**, 533–546.

MOREMAN, W. L. 1942. Paleontology of the Eagle Ford Group of north and central Texas. *Journal of Paleontology*, **16**, 199–220.

MÜLLER, G. 1888. Beitrag zur Kenntnis der oberen Kreide am nördlichen Harzrande. *Jahrbuch des Preussischen Geologischen Landesamts*, **8**, 372–456, pls 16–18.

NAGAO, T. and MATSUMOTO, T. 1939. A monograph of the Cretaceous *Inoceramus* of Japan. Part I. *Journal of the Faculty of Science, Hokkaido Imperial University, Series 4*, **4**, 241–299, pls 23–34.

—— —— 1940. A monograph of the Cretaceous *Inoceramus* of Japan. Part II. *Journal of the Faculty of Science, Hokkaido Imperial University, Series 4*, **6**, 1–64, pls 1–22.

NIEBUHR, B., BALDSCHUHN, R., ERNST, E., WALASZCZYK, I., WEISS, W. and WOOD, C. J. 1999. The Upper Cretaceous succession (Cenomanian–Santonian) of the Staffhorst Shaft, Lower Saxony, northern Germany: integrated biostratigraphic, lithostratigraphic and downhole geophysical log data. *Acta Geologica Polonica*, **49**, 175–213.

NODA, M. 1975. Succession of *Inoceramus* in the Upper Cretaceous of southwest Japan. *Memoirs of the Faculty of Science, Kyushu University, Series D, Geology*, **23**, 211–261, pls 32–37.

—— 1984. Notes on *Mytiloides incertus* (Cretaceous Bivalvia) from the Upper Turonian of the Pombets Area, central Hokkaido. *Transactions and Proceedings of the Palaeontological Society of Japan, New Series*, **136**, 455–473, pls 84–86.

—— 1996. Five inoceramids (Bivalvia) from the Upper Cretaceous of Hokkaido with some phylogenetic and taxonomic considerations. *Transactions and Proceedings of the Palaeontological Society of Japan, New Series*, **184**, 555–591.

—— and MATSUMOTO, T. 1999. Palaeontology and stratigraphy of the inoceramid species from the mid-Turonian through upper Middle Coniacian in Japan. *Acta Geologica Polonica*, **48** (for 1998), 435–482, pls 1–18.

—— and MURAMOTO, K. 1980. A new species of *Inoceramus* (Bivalvia) from the Upper Cretaceous of Hokkaido. *Transactions and Proceedings of the Palaeontological Society of Japan, New Series*, **119**, 388–402.

—— and TOSHIMITSU, S. 1990. Notes on a Cretaceous bivalve *Inoceramus* (*Platyceramus*) *mantelli* de Mercey from Japan. *Transactions and Proceedings of the Palaeontological Society of Japan, New Series*, **158**, 485–512.

NUMMEDAL, D. and MOLENAAR, C. M. 1995. Sequence stratigraphy of ramp-setting strand plain successions: the Gallup Sandstone, New Mexico. *In* VAN WAGONER, J. C. and BERTRAM, G. T. (eds). Sequence stratigraphy of foreland basin deposits – outcrop and subsurface examples from the Cretaceous of North America. *Memoir of the American Association of Petroleum Geologists*, **64**, 277–310.

ORBIGNY, A. D' 1843–1847. *Paléontologie Francaise, terrains Crétacés 3*. Masson & Cie, Paris, 807 pp.

PARKINSON, J. 1818. Remarks on the fossils collected by Mr. Phillips near Dover and Folkstone. *Transactions of the Geological Society, London*, **5**, 1–55.

PAULIUC, S. 1968. Studiul geologic al Perşanilor centrali cu privire specială la Cretacicul superior. *Studii Tehnice şi Economice, Serie J, Stratigrafie*, **4**, 1–133, 33 pls.

PAVLOVA, M. M. 1955. Inoceramids of the Upper Cretaceous deposits of Daghestan. Unpublished PhD thesis, Moscow State University, Moscow. [In Russian].

PERGAMENT, M. A. 1967. Stages in *Inoceramus* evolution in the light of absolute geochronology. *Paleontological Journal*, **1**, 27–34. [AGI translation].

—— 1971. Biostratigraphy and inocerams of Turonian–Coniacian deposits of the Pacific regions of the USSR. *Transactions of the Academy of Sciences of the USSR, Geological Institute*, **212**, 1–196, pls 1–73. [In Russian].

—— 1978. Upper Cretaceous stratigraphy and inocerams of the Northern Hemisphere. *Transactions of the Academy of Sciences of the USSR, Geological Institute*, **322**, 1–214. [In Russian].

PETRASCHECK, W. 1903. Ueber Inoceramen aus der Kreide Böhmens und Sachsen. *Jahrbuch der Kaiserlich-Königlichen Reichsanstalt*, **53**, 153–168, pl. 8.

RADWAŃSKA, Z. 1962. The fauna of the bottom beds of *Inoceramus schloenbachi* Zone from Wilkanów (Lower Silesia). *Biuletyn Instytutu Geologicznego*, **173**, 129–167, pls 1–8. [In Polish, English summary].

—— 1963. Die Grenze zwischen dem Turon und dem Coniac in der Innesudetiscehn Mulde und im Neissegraben. *Berichte der Geologischen Gesselschaft der DDR*, **8**, 163–170, pls 5–6.

RANKIN, C. H. 1944. Stratigraphy of the Colorado Group, Upper Cretaceous, in northern New Mexico. *Bulletin of the New Mexico School of Mines*, **20**, 1–27.

RAWSON, P. F., DHONDT, A. V., HANCOCK, J. M. and KENNEDY, W. J. 1996. Proceedings, Second International Symposium on Cretaceous Stage Boundaries, Brussels 8–16 September, 1995. *Bulletin de l'Institut Royal des Sciences Naturelles de Belgique*, **66** (Supplement), 1–117.

REESIDE, J. B. Jr 1923. A new fauna from the Colorado Group of southern Montana. *Professional Paper of the United States Geological Survey*, **132-B**, 25–33, pls 45–50.

—— 1927. The Scaphites, an Upper Cretaceous ammonite group. *Professional Paper of the United States Geological Survey*, **150-B**, 21–40, pls 9–11.

—— 1929. The Cretaceous section in Black Mesa, northeastern Arizona. *Journal of the Washington Academy of Sciences*, **19**, 30–37.

—— 1930. The Cretaceous faunas in the section on Vermilion Creek, Moffat County, Colorado. *Journal of the Washington Academy of Sciences*, **20**, 35–41.

ROBINSON, C. S., MAPEL, W. J. and BERGENDAHL, M. H. 1964. Stratigraphy and structure of the northern and western flanks of the Black Hills uplift, Wyoming, Montana, and South Dakota. *Professional Paper of the United States Geological Survey*, **404**, 1–134.

RUBEY, W. W. and BASS, N. W. 1925. The geology of Russell County, Kansas, with special reference to oil and gas resources. *Bulletin of the Kansas Geological Survey*, **10**, 1–86.

RYER, T. A. 1976. Cretaceous stratigraphy of the Coalville and rockport areas, Utah. *Utah Geology*, **3**, 71–83.

—— 1977*a*. Patterns of Cretaceous shallow-marine sedimentation, Coalville and Rockport areas, Utah. *Bulletin of the Geological Society of America*, **88**, 177–188.

—— 1977*b*. Coalville and Rockport areas, Utah. *The Mountain Geologist*, **14**, 105–128.

SALVADOR, A. (ed.). 1994. *International stratigraphic guide; a guide to stratigraphic classification, terminology, and procedure. Second Edition*. International Union of Geological Sciences and Geological Society of America, Boulder, Colorado, 214 pp.

SCHLÜTER, C. 1877. Kreide-Bivalven. Zur Gattung *Inoceramus*. *Palaeontographica*, **24**, 250–288, pls (1)36–(4)39.

SCOTT, G. R. 1986. Geologic and structure contour map of the Springer 39' × 60' quadrangle, Colfax, Harding, Mora, and Union Counties, New Mexico. *United States Geological Survey, Miscellaneous Investigations Series, Map* **I-1705**, scale 1:100,000.

—— and COBBAN, W. A. 1964. Stratigraphy of the Niobrara Formation at Pueblo, Colorado. *Professional Paper of the United States Geological Survey*, **454-L**, 1–30, pls 1–11.

—— —— and MEREWETHER, E. A. 1986. Stratigraphy of the Upper Cretaceous Niobrara Formation in the Raton Basin, New Mexico. *Bulletin of the New Mexico Bureau of Mines and Mineral Resources*, **115**, 5–34.

SCUPIN, H. 1912–1913. Die Löwenberger Kreide und ihre Fauna. *Palaeontographica*, **6**, 1–276, pls 1–15.

SEIBERTZ, E. 1979. Biostratigraphie im Turon des SE-Münsterlandes und Anpassung an die internationale Gliederung aufgrund von Vergleichen mit anderen Oberkreide-Gebieten. *Newsletters on Stratigraphy*, **8**, 111–123.

—— 1986. Paleogeography of the San Felipe Formation (mid-Cretaceous, NE Mexico) and facial effects upon the inoceramids of the Turonian/Coniacian transition. *Zentralblatt für Geologie und Paläontologie*, **9/10**, 1171–1181.

SEITZ, O. 1959. Vergleichende Stratigraphie der Oberkreide in Deutschland und in Nordamerika mit hilfe der Inoceramen. *Congreso Geologico Internacional. XX Sesion, Ciuadao de Mexico, 1956*, 113–129.

—— 1965. Die Inoceramen des Santon und Unter-Campan von Nordwestdeutschland. II. Teil. (Biometrie, Dimorphismus und Stratigraphie der Untergattung *Sphenoceramus* J. Böhm). *Beihefte der Geologischen Jahrbuch*, **69**, 1–194.

—— 1967. Die Inoceramen des Santon und Unter-Campan von Nordwestdeutschland, Teil III, Taxonomie und Stratigraphie der Untergattungen *Endocostea, Haenleinia, Platyceramus, Cladoceramus, Selenoceramus* and *Cordiceramus* mit besonderer Berücksichtigung des Parasitismus bei disen Untergattungen. *Beihefte der Geologischen Jahrbuch*, **75**, 1–171.

SHIMER, H. W. and SHROCK, R. R. 1944. *Index fossils of North America*. John Wiley, New York, 837 pp., 303 pls.

SIMIONESCU, J. 1899. Fauna cretacică superióră de la Ürmös (Transilvania). *Academia Română, Publicaţiunile Fondului Vasilie Adamachi*, **4**, 238–274.

SORNAY, J. 1974. Inocérames Turoniens d'Afghanistan. *Annales de Paléontologie*, **60**, 27–34.

—— 1980. Révision du sous-genre d'Inocérame *Tethyoceramus* Heinz 1932 (Bivalvia) et de ses représentants Coniaciens a Madagascar. *Annales de Paléontologie*, **66**, 135–150.

STANTON, T. W. 1894. The Colorado Formation and its invertebrate fauna. *Bulletin of the United States Geological Survey*, **106**, 3–189, 45 pls [1893 imprint].

STOSE, G. W. 1932. *Geologic map of the United States*. Scale 1:2,500,000. United States Geological Survey.

SZÁSZ, L. 1985. Contributions to the study of the *Inoceramus* fauna of Romania. I. Coniacian *Inoceramus* from the Babadag Basin (North Dobrogea). *Memoriile Institutului de si Geophysica* **32**, 137–184, pls 1–40.

TARKOWSKI, R. 1991. Stratigraphy, macrofossils and palaeogeography of the Upper Cretaceous from the Opole Trough. *Geology, Scientific Bulletins of Stanislaw Staszic Academy of Mining and Metalurgy*, **51**, 1–156, 28 pls.

TOURTELOT, H. A. and COBBAN, W. A. 1968. Stratigraphic significance and petrology of phosphate nodules at base of Niobrara Formation, east flank of Black Hills, South Dakota. *Professional Paper of the United States Geological Survey*, **594-L**, 1–22.

TREXLER, D. W. 1966. Stratigraphy and structure of the Coalville area, Utah. *Professional Contributions of the Colorado School of Mines*, **2**, 1–69.

TRÖGER, K.-A. 1967. Zur Paläontologie, Biostratigraphie und faziellen Ausbildung der unteren Oberkreide (Cenoman bis Turon). Teil I: Paläontologie und Biostratigraphie Inoceramen des Cenomans bis Turons. *Abhandlungen des Staatlichen Museums für Mineralogie und Geologie zu Dresden*, **12,** 13–208, pls 1–14.

—— 1974. Zur Biostratigraphie des Ober-Turon bis Unter-Santon aus dem Schachtaufschluss der Zeche Grimberg IV bei Bergkamen. *Freiberger Forschungshefte*, **C245**, 68–81.

—— 1976. Evolutionary trends of Upper Cretaceous inocerames. *Evolutionary Biology* (Praha), 193–203.

—— 1981*a*. Zur Bedeutung der Wachstumsknicke bei Inoceramen der Oberkreide. *Freiberger Forschungshefte*, **C363**, 101–110.

—— 1981*b*. Zu Problemen der Biostratigraphie der Inoceramen und der Untergliederung des Cenomans und Turons in Mittel- und Osteuropa. *Newsletters on Stratigraphy*, **9**, 139–156.

—— 1985. *Inoceramus* (*Mytiloides*) *incertus prescheri* n. ssp., eine neue Unterart der *Inoceramus incertus*-Gruppe us dem Oberturon von Dresden-Strehlen. *Freiberger Forschungshefte*, **C410**, 41–44.

—— 1989. Problems of Upper Cretaceous inoceramid biostratigraphy and paleobiogeography in Europe and Western Asia. 911–930. *In* WIEDMANN, J. (ed.). *Cretaceous of the Western Tethys.* E. Schweitzerbart'sche, Stuttgart, 1005 pp.

—— and CHRISTENSEN, W. K. 1988. Upper Cretaceous (Cenomanian–Santonian) inoceramid bivalve faunas from the island of Bornholm, Denmark. *Danmarks Geologiske Undersøgelse, Serie A*, **28**, 1–47.

—— and SUMMESBERGER, H. 1994. Coniacian and Santonian bivalves from the Gosau-Group (Cretaceous, Austria) and their biostratigraphic and palaeobiogeographic significance. *Annalen des Naturhistorischen Museum in Wien*, **96A**, 161–197.

TWETO, O. 1979. *Geologic map of Colorado*. Scale 1:500,000. United States Geological Survey.

VOIGT, S. 1995. Palaeobiogeography of early Late Cretaceous inoceramids in the context of a new global palaeogeography. *Cretaceous Research*, **16**, 343–356.

—— and HILBRECHT, H. 1997. Late Cretaceous carbon isotope stratigraphy in Europe: correlation and relations with sea level and sediment stability. *Palaeogeography, Palaeoclimatology, Palaeoecology*, **134**, 39–59.

WALASZCZYK, I. 1992. Turonian through Santonian deposits of the Central Polish Uplands; their facies development, inoceramid paleontology and stratigraphy. *Acta Geologica Polonica,* **2**, 1–122, pls 1–42.

—— 1996. Inoceramids from Kreibitz-Zittauer area (Saxony and northern Bohemia): revision of Andert's (1911) description. *Paläontologische Zeitschrift*, **68**, 367–392.

—— 1997. Significance of the ligament area in species level taxonomy of inoceramid bivalves; how much variation is lodged in a single species? *Freiberger Forschungshefte*, **C468**, 289–303.

—— and COBBAN, W. A. 1999. The Turonian–Coniacian boundary in the United States Western Interior. *Acta Geologica Polonica*, **48** (for 1998), 495–507.

—— and SZÁSZ, L. 1997. Inoceramid bivalves from the Turonian/Coniacian (Cretaceous) boundary in Romania: revision of Simionescu's (1899) material from Ürmös (Ormenis), Transylvania. *Cretaceous Research*, **18**, 767–787.

—— and TRÖGER, K.-A. 1996. The species *Inoceramus frechi* (Bivalvia, Cretaceous); its characteristics, formal status, and stratigraphic position. *Paläontologische Zeitschrift*, **68**, 393–404.

—— and WOOD, C. J. 1999*a*. Inoceramids and biostratigraphy at the Turonian/Coniacian boundary; based on the Salzgitter-Salder (proposed boundary stratotype) section (Lower Saxony, Germany), and the Słupia Nadbrzeżna section (central Poland). *Acta Geologica Polonica*, **48**, 395–434, pls 1–19.

—— —— (eds) 1999*b*. Inoceramid record and biostratigraphy across the Turonian/Coniacian boundary & report on the Second Inoceramid Workshop, Freiberg, 1996. *Acta Geologica Polonica* **48**, i-iv.

WIESE, F. 1997. Das Turon und Unter-Coniac im Nordkantabrischen Becken (Provinz Kantabrien, Nordspanien): Faziesentwicklung, Bio-, Event- und Sequenzstratigraphie. *Berliner Geowissenschaftliche Abhandlungen, E*, **24**, 1–131.

WHITE, C. A. 1874. Preliminary report upon invertebrate fossils collected by the expeditions of 1871, 1872, and 1873. *United States Army Engineering Department, Geographical and Geological Explorations and Surveys West of the 100th Meridian*, 1–27.

—— 1877. Report upon the Cretaceous invertebrate fossils collected in portions of Nevada, Utah, Colorado, new Mexico, and Arizona, by the parties of the expeditions of 1871, 1872, 1873, and 1874. *United States Geographical and Geological Explorations and Surveys West of the 100th Meridian, Report*, **4** (1), 1–219.

—— 1878. Report on the geology of a portion of northwestern Colorado. *United States Geological and Geographical Survey of the Territories (Hayden), Annual Report*, **10**, 1–60.

—— 1879. Contributions to invertebrate paleontology, No. 1: Cretaceous fossils of the western States and Territories. Department of the Interior. *United States Geological Survey, Eleventh Annual Report of the Survey for the year 1877*, 273–319.

WHITFIELD, R. P. 1877. Preliminary report on the paleontology of the Black Hills, containing descriptions of new species of fossils from the Potsdam, Jurassic, and Cretaceous formations of the Black Hills of Dakota. *United States Geographical and Geological Survey of the Rocky Mountain Region*, 1–49.

—— 1880. Paleontology of the Black Hills of Dakota. *In* NEWTON, H. and JENNEY, W. P., Report on the geology and resources of the Black Hills of Dakota. *United States Geographical and Geological Survey of the Rocky Mountain Region*, 325–468.

WIEDMANN, J. and KAUFFMAN, E. G. 1978. Mid-Cretaceous biostratigraphy of northern Spain. *Annales du Museum d'Histoire Naturelle de Nice*, **4**, III1–III34.

WOOD, C. J., ERNST, G. and RASEMANN, G. 1984. The Turonian-Coniacian stage boundary in Lower Saxony (Germany) and adjacent areas: the Saltzgitter-Salder Quarry as a proposed international standard section. *Bulletin of the Geological Society of Denmark*, **33**, 225–238.

WOODS, H. 1912. A monograph of the Cretaceous Lamellibranchia of England. Volume 2, Part 8. *Monographs of the Palaeontographical Society* (for 1911), 285–340, pls 51–54.

YEHARA, S. 1924. On the Izumi sandstone group in the Onogawa basin, Prov. Bungo and the same group in Uwajima, Prov. Iyo. *Japanese Journal of Geology and Geography*, **3**, 24–40.

ZONOVA, T. D. 1970. Upper Cretaceous inoceramids of the *Inoceramus uwajimensis* group and their stratigraphical meaning. *Transactions of the All-Union Geological Institute, VSEGEI, New Series*, **127**, 174–200. [In Russian].

—— 1992. Cretaceous Inoceramids of eastern USSR. 172–191. *In* ZONOVA, T. D. and ROSTOVTSEV, K. O. (eds). Atlas of the Mesozoic guide fossil faunal groups of southern and eastern USSR. *Transactions of VSEGEI, New Series*, **350**, 375 pp. [In Russian].

—— and YAZYKOVA, E. A. 1999. Biostratigraphy and correlation of the Turonian–Coniacian succession and the Turonian-Coniacian boundary problem in the Far East Russia based on ammonites and inoceramids. *Acta Geologica Polonica*, **48**, 483–494, pls 1–14.

I. WALASZCZYK

Institute of Geology
University of Warsaw
Al. Zwirki i Wigury 93
PL-02-089 Warszawa, Poland
e-mail walas@geo.uw.edu.pl

W. A. COBBAN
70 Estes Street
Lakewood
Colorado 80226, USA

Typescript received 5 July 1999
Revised typescript received 31 May 2000